Media, Technology and the Imagination

Media, Technology and the Imagination

Edited by

Marie Hendry and Jennifer Page

CAMBRIDGE SCHOLARS

PUBLISHING

Media, Technology and the Imagination,
Edited by Marie Hendry and Jennifer Page

This book first published 2013

Cambridge Scholars Publishing

12 Back Chapman Street, Newcastle upon Tyne, NE6 2XX, UK

British Library Cataloguing in Publication Data
A catalogue record for this book is available from the British Library

ISBN (10): 1-4438-4850-6, ISBN (13): 978-1-4438-4850-3

This collection is dedicated in loving memory to Dr. Barbara J. Cicardo for her continued support as a mentor and to the conference on Literature, Language and Culture. “Bobbie C.” will be sorely missed.

TABLE OF CONTENTS

ACKNOWLEDGEMENTS

The editors of this collection wish to thank all those that have supported and continue to support the conference on Literature, Language and Culture at the University of Louisiana at Lafayette. In particular, special thanks go to Dr. Chris Healy and Dr. Clai Rice for their support during the March 2012 conference. Also, we wish to thank Dr. Christine DeVine for her work in previous conference years and to the Department of English at the University of Louisiana at Lafayette. We would also wish to thank Joel T. Terranova and Laura Holder for their help in peer-reviewing and editing this volume.

The editors also wish to thank everyone at Cambridge Scholars Publishing for their help and support of this volume.

INTRODUCTION

MARIE HENDRY AND JENNIFER PAGE

The essays collected in this volume reflect the scholarly conversation started at the eleventh annual Louisiana Conference on Literature, Language and Culture, held at the University of Louisiana at Lafayette in March 2012. Because this conference continually welcomes inquiry from academics at all stages of development and from a multiplicity of disciplines, the theme of media, technology and the imagination was especially enticing and provocative. Scholarly advancement now hinges upon technology in ways that were previously unimagined. Contemporary analysts might take for granted that technologies like word processing software, online research databases, discussion boards, and listservs, as quotidian as they now are, have irreversibly changed the standard method of performing literary and cultural analysis. As a symbol of literary innovation and query, the quill seems to have been replaced by the blinking cursor. Certainly, the intersection of imaginative thought and technology is beneficial for scholars, especially as it allows us to share discoveries and new ways of thinking; media and technology, for the modern thinker, foster the development of the imagination.

Beyond the typical fodder of literary analysis—which itself must acknowledge that technological innovation is commonplace in contemporary literature, no longer an element of only speculative or science fiction—a discussion of media, technology and the imagination invited speculation on generic performance, narrative film, and documentaries as well. Because of Louisiana's rich performance history, as the birthplace of jazz and the home to the zydeco musical genre, and as an attractive setting for Hollywood productions, from the well-lauded *Beasts of the Southern Wild* to the wildly popular *Twilight Saga* franchise, an investigation of the crux of media and the imagination was surprisingly well-suited for French Acadiana. Moreover, the newly-established Moving Image Arts program at UL Lafayette, orchestrated by the documentarian and humanitarian Charles E. Richard, generated further interest in this mutli-genre, multi-cultural conversation. With the support of local filmmakers and producers, the Louisiana Conference welcomed the opportunity to screen documentaries and host intimate discussions about using film to convey individual imagination.

As deeply as media and technology influence research and writing habits, they also play an important role in our professional development and pedagogical strategies, and several contributors to the Louisiana Conference questioned how technology can be used to enrich students' learning experiences in the traditional and online classroom. One might think of online discussion boards or database research as some of the main methods of urging student familiarity with technology; however, instructors in the humanities are incorporating digital tools to demonstrate that artistic and cultural inquisition can be methodical as well as intuitive. Through the use of blogs, collaborative wikis, as well as software developed for visual analysis, desktop publishing, and film editing, instructors are introducing students to technologies that allow them to develop their imagination and present themselves as impressively tech-savvy. Because several presenters shared their knowledge and experiences with introducing students to new technology, other researchers at the conferences will be able to integrate methods and technologies into their own courses.

It is fitting that a conference so invested in interdisciplinary exchange and growth included the unique perspective of a scholar whose corpus embodies the paradoxical, inspiring marriage of technology and the literary imagination. As a geneticist, administrator, playwright, and poet, John Doucet's essay—located at the beginning of this volume—celebrates the symbiotic relationship of technological and imaginative innovation. Part autobiography, part ode to comic book art, and part exploration and condemnation of the arbitrary delineation of educational pursuits, Doucet's essay affirms that technology and artistry are not mutually exclusive; rather, they augment one another, leaving researchers, students, and artists with a nearly limitless potential of invention and investigation.

The essays in this collection likewise investigate more specific ways in which media, technology and the imagination inspire dynamic intellectual pursuits. Some of the essays take classical literary texts and writers, such as Milton, and discuss how technology has given these works a new birth. Others deal with modern texts and television, such as Laura Holder's ending essay on the television show *Buffy, The Vampire Slayer* and traditional monster motifs.

It is our hope that this collection will inspire readers to reflect on the role(s) of pop-culture in academia and the multi-facetedness of how neo-studies produce a unique interpretation of media and technology. Most importantly, it is the imaginative nature of this volume that stands as a testament for the importance of this theme in today's emerging dialogue on pop-culture.

PART I:

PERSONAL ESSAYS ON SPECTATING AND BECOMING

CHAPTER ONE

ON THE DESIGN OF MENTAL ORGANISMS

JOHN P. DOUCET[1]

It is the nature of mental organisms to resolve thoughts, concepts, and ideas to conclusion. A conclusion can be logical, based on structures observed from the sensuous world and chemically remanufactured as a series of neurotransmissions across synapses of neurons that may be recalled time and time again in the process of memory and then used as rationale to explain other, unknown thoughts, concepts, and ideas theretofore unconcluded. Alternatively, a conclusion can be illogical, which would be nonetheless a similar chemically remanufactured observation that can be recalled and then used by casual and intentional observers alike as rationale for dirty looks, snide remarks, lack of career promotion, and expensive prescriptions with unwelcomed side effects.

A cat, for example, is certainly a mental organism, as it remanufactures the location of its supper bowl and time of filling on multiple occasions per day. Its conclusion is that this location is where sustenance will be provided by larger mental organisms who routinely fail to realize that it can provide sustenance for itself by hunting for juicier kibbles, like mice. The cat at its location, however, will not make the conclusion that its supper bowl will be filled by a mouse slowly and involuntarily crawling across the surface of its supper bowl feigning its traditional role as prey. That would be an illogical conclusion, and cats do not make these.[2]

Lions are another example. Despite also being considered a "cat," lions have clear differences in their acumen for making conclusions about the

[1] The author is a mild-mannered scientist rendered superhero by mutation—presumably not mutation induced by cosmic rays, gamma rays, or the bite of a radioactive spider but rather by the natural, spontaneous mutations that occur in the stem cells of parental germlines. At least that's what his meager superpowers suggest. For more on germlines mutations, see note 46.

[2] Evidence for this is, of course, anecdotal, and is sometimes challenged by other evidence provided by YouTube and *America's Funniest Videos*.

natural world. For one thing, they would not even conjure the thought of a mouse as food, preferring the more slender and tender dark meat of the wildebeest or gazelle.[3] They would also not depend on larger mental organisms to provide sustenance, for, oppositely, larger mental organisms sometimes actually serve as sustenance for the lion itself. Most importantly, and unlike the domesticated cat, a much smaller mental organism, lions do make the occasional slight of logic—like trying to paw down two rhinos at once or trying to swallow that first bite of alligator without first removing the skin. Lions never remanufacture the series of neurotransmissions necessary to correct this illogic because no other mental organism—not humans, not mice—will exert the requisite fortitude to alert the larger, carnivorous mental organisms to their failures. Fortitude, it seems, is a prelude to face-to-face criticism, and in its conclusion-making the lion has no immediate need to recognize that behavior in other mental organisms.[4]

Humans, on the other hand[5], are mental organisms distinct from cats large and small in that in humans it is insufficient simply to resolve to conclusion. Other, confounding synapses manufacture the *desire* to come to conclusion, to covet *meaning,* and to use meaning to generate more complex thoughts and concepts and ideas.

One manifestation of this uniquely human desire for conclusion is the academic conference. Attended mostly by humans with the occasional post-conference gathering of smaller cats around garbage cans where the leftover refreshments not otherwise absconded by graduate students are left for refuse, such conferences invite an array of specialized thoughts, concepts, and ideas with an inherent understanding among participants that a conclusion will be resolved. This resolution usually manifests in the spectacular ideas of one invited human with handsome gifts of both complexity of thought and eloquence of word.

And so, I begin.

Of all the scholastic, convivial and collegial motivations to undertake this conference, perhaps none has provided the requisite "ire" in our *desire* as mental organisms to come to conclusion as one motivation in particular. Our contemporary, literary theorist, and critic Terry Eagleton, has provided many intellectual challenges to address at many academic conferences.

---

[3] That's "legs and thighs" for you eaters of potted beef tripe and mechanically-separated chicken, even if it's canned as "America's Favorite."

[4] With the exceptions, of course, of moments under contractual obligations to circuses, zoos, and most Siegfried and Roy performances. At one point or another, all mental organisms work for meat and money.

[5] Paw?

This conference, on media, technology, and the imagination, has chosen its own challenge from his book, *Literary Theory*: "Superman comic and Mills and Boon novels are ... not generally regarded as literature, and certainly not Literature."[6]

And so, at the end of this conference, I am left with a multitude of ends to tie, a plethora of parallel ports if you technologically will, including lyrics by both Iggy Pop and Terry Eagleton, none of which can be sung by the human voice and maybe not even that of the cat.[7] By way of resolving "media, technology and the imagination" to conclusion, I here offer a graphic, novel, largely autobiographical apology in defense of our collective ideas.

Megacephalization

Evolutionary biologists explain that cephalization was the consequence of natural selection in the prehistoric struggle to find food. First, there was some sort of cell or body structure in an ancestral organism containing a concentration of pigment molecules that could receive and respond to light photons and thereby distinguish dark—like distinguishing the silhouette of something tasty swimming across a light source. It was beneficial to organisms to have these pigmentary sensors developed near the organismal mouth where the tasties would be captured. Those with more pigment molecules experienced better acuity and were fitter to eat, survive, mate, and distribute their genes to a next generation. More pigment molecules meant more and bigger cells, bigger tissues, and then bigger heads.

This is how evolutionary biologists explain cephalization. There are some things, however, that even the most brazen of such scientists fear to explain.

* * *

[6] Eagleton, Terry. *Literary Theory: An Introduction*. Minneapolis, MN: University of Minnesota Press, 1983. Print. P. 2. (Hereafter, *Literary Theory*.) Readers of *Literary Theory* may recognize that our never illogical "cat" is also the favorite, furry signified referent of Ferdinand de Saussure and semiotics (pp. 96-7).

[7] This is only a supposition. During some seasonal cat fights, I have heard harmonic approximations of Iggy Pop melodies. Although I have heard none of Eagleton's during these episodes, I suspect this is because those longer-winded cats are routinely eliminated in the natural selection inherent to the fight.

It was a harrowing diagnosis, one that a doctor should never reveal to a mother.[8] But this was the 1960s, a time when lawyers were busy solving the Kennedy assassination and not concerned about emotional traumas induced by careless or insensitive physicians.

Megacephaly. A large head. Of all things to be large. A father inevitably would be prouder if the diagnosis applied to something other than the cranium. But I speak in generalities. My father was proud, but perhaps this was because he couldn't see *past* the cranium.

My mother has always been proud. Or maybe she was proud just to have survived the natural birth of a megalo.[9] That earned her membership in an exclusive club in the world of women, and maybe she went home with special gifts, like not one but two boxes of those new-fangled disposable diapers and not one but two cases of that experimental milk formula. I believe that there were not one but two sheets of S&H Green Stamps thrown in as a consolation prize, but I could be mistaken. After all, I was young.

A few years after birth, and after missing a few developmental landmarks, I had come to realize that I was not clinically megacephalic.[10] But I still had a big head.[11] And that would prove a big fricative problem.

It's hard to be a big kid. It's worse to be a big-headed kid.

For one thing, big-headed kids have big head colds in winter because there's so much surface area to catch a chill.[12] And, on windy days, forget finding a cap or a hat that fits. My mother once tried to buy a cap for me, and the lady at the register called the produce section for a price check. Another lady asked if I came seedless.[13]

---

[8] Of course, since she had already birthed the subject, perhaps there was suspicion among the doctors that she was already desensitized to the situation.

[9] I sorta helped out—at least initially. I was breech.

[10] For instance, I learned to walk later than the accepted pediatric standard age because I was perfectly adept at getting from one place to another by rolling on my head.

[11] So, that would make me *macrocephalic,* not *megacephalic* or worse—*megalencephalic*. But, a curse is a curse by any other name. I think Shakespeare said that, maybe in *Macbeth.*

[12] Now, doctors will tell you that colds are caused by viruses and not by frigid, ambient temperature. Others will tell you that frigid, ambient temperatures lower your immune response, making you more susceptible to viruses. Either way, I still had a big head and big head colds.

[13] That lady may have had an ulterior motive, as cougars, by definition, often do. Pardon the cat analogy: "Thrice [now] the brinded cat hath mewed" (Shakespeare, William. *The Tragedy of Macbeth. The Norton Shakespeare.* Ed. Stephen Greenblatt.

Sports were a disaster. I made the local Biddy Basketball team not as a player but as equipment. For football, I couldn't fit any helmets, so I was converted to place kicker.[14] When daddies would take their sons to see a baseball doubleheader, they'd end up at my house to sit and watch me.

Elementary school itself was not much more fun. I wouldn't play hide-and-seek because my head or its large shadow would inevitably give me away. I wouldn't play with hula-hoops because they'd painfully pull my hair when I'd try to put them on. And the only reason I was ever invited to swimming parties is because parents viewed me as a safe flotation device.

I remember going to the cinema as a kid. People complained even though I sat in the back row. If you ever wonder who invented stadium seating, it's the guy who was sitting behind me during *Star Wars*.[15]

Speaking of outer space: Once, after a lunar eclipse over my hometown, people would call my parents and ask me to go out and do it again. One day in second grade, we got to act out the solar system. All the kids said I should be Jupiter. "No," Mrs. Sensitivity said, "He has to play the sun because his head is so [whatever]." At least I was the center of planetary attention. And, though I secretly wanted to be a "star" pupil, this was not what I had envisioned.[16]

Some autumn mornings as a kid, I'd wake up to a bed full of acorns. My three elder sisters would tell me that the backyard squirrels needed help transporting large volumes of these oak nuts for winter. And so, they explained, during the night the furry animals would silently bring me a large stash to carry in my cheeks to the burial grounds. How my sisters knew details of this innate squirrel behavior was a mystery.

The emotional trauma for a kid with a big head has long-lasting effects. For years I had this recurring dream that I was at a birthday party in Mexico surrounded by children carrying sticks who were crying

---

2nd edition. New York: W. W. Norton & Co., 2008. 2579-2632. IV.i.1. Hereafter, *Macbeth*.

[14] I am being polite. This was really a demotion. Kickers practiced separately, used separate showers, rode separate buses for away games, and enjoyed only the glory of being the default scorer when the much preferred touchdown could not be achieved by the more lauded players of the offense who have below normal-to-normal cranial diameters.

[15] Apparently, and unbeknownst to me except for the slight scalp burn I suffered at the back of my head, I had eclipsed close-up scenes of the Death Star several times.

[16] In addition, as the sun and not as Jupiter, I got to stand much closer to the precocious blonde girl who played Venus, a consolation I didn't fully appreciate until several years later when I approached high school years. (See note 26.)

inconsolably because they couldn't beat candy out of my head. Another time I dreamed I was swimming somewhere off the Gulf Coast and Gregory Peck tried to harpoon me.[17] But the worst dream I ever had was when I imagined my head was infested with lice sailing in tiny wooden sailing boats. I heard one louse exclaim in Spanish, "It's flat, Christopher! For God's sake, man! Turn back!"[18]

Orwellian Proportions[19]

It is a great day in high school when a student earns reprise from his teacher's classroom rantings and instead gets to watch a movie. That is, such is a great day in high school if you have a normal sized head. By those days, however, the non-cranial aspects of my body had normalized, at least somewhat so, and thus the cranium itself was less conspicuous given that growth phenomenon, together with emerging expression of unexpected personal charm previously concealed in those of my more disproportionate elementary school days.

Beyond the typical directions of my teachers to make sure that I sit in the back of the class, which became so commonplace as to be no longer embarrassing or discouraging, it's always a risk for a big-headed kid because particular items in films—like the Death Star of *Star Wars*—that can alert detractors to the shape and unappreciated grandeur of his cranial disposition.

But there is no day **less** such a great day than that unforgotten day when my English teacher wheeled in a vacuum-tube television monitor, which sat atop a three-fourths-inch video tape player on a four-foot tall

---

[17] This Technicolor dream was a leftover memory of watching the 1956 Warner Brothers film *Moby Dick*, which starred Peck. I survived that dream by blasting Peck and his feeble harpoon apparatus with a great snort from my dorsal blowhole, no doubt a Freudian concoction deriving from many years as a tuba player in various school bands.

[18] Maybe it was Portuguese. It's hard to tell when you're sleeping.

[19] "Orwellian" in this usage is an original contraction of "Orson" and "Welles," after the macrocephalic director, play-adaptor, and titular star of the film *Macbeth*, which is the subject of ensuing paragraphs. The term at this point, much like Eagleton's *Literary Theory*, has little to do with George Orwell himself, despite that author's importance to all of literature in, among other things, illustrating the idea and consequence of a dystopian world, a place where, it is abundantly apparent from his writings, the critic Eagleton lives.

cart[20] and told us that we were about to watch *Macbeth*.[21] This was not just some friendly caught-live-on-film play between actors mouthing obscure, overwritten lines with funny accents. This was the famous[22], stark, spare, 1948 Republic Pictures version for which Orson Welles was awarded only $700,000 to headline and produce and which he completed in only twenty-three days.[23]

Of course, the script was Shakespeare's, and so there could be no criticism of or detraction from that exalted regard in the world of public school English class.[24] And, as prelude, the class was warned of the "curse of Macbeth"—the injuries, riots, and deaths that had occurred through history at theaters where Macbeth was performed. This frightened us deeply. As if we were in imminent danger, we were uncharacteristically hushed.[25]

As the film began and the stark, black-and-white images cemented all our fears, it became clear that the imminent danger was all and only mine. In the first two minutes of the film, two realizations ignited inside that great vault where my self-consciousness paced ceaselessly and sent fire down my spine and into my skin to flush with embarrassment. These two anxieties put me once again at risk of ridicule by classmates and the

---

[20] Okay, this was the seventies—the *late* seventies to be exact. Nothing digital, nothing streaming, nothing in current usage, all obsolete. But, to paraphrase Iggy Pop in the roughly contemporaneous proto-punk anthem "Search and Destroy," "Ain't got time to make no apology! Look out, honey, we're using technology—along with media and imagination!" (Iggy and the Stooges. "Search and Destroy." *Raw Power*. Record album. CBS. 1973.) Hereafter, "Search and Destroy." In the first line of this song, among a litany of self-descriptions, Iggy describes himself as a "street-walking cheetah"—yet another brinded mew. It's a connecting principle. It's synchronicity, man. Somebody call Carl Jung (Jung, Carl. *Synchronicity: An Acausal Connecting Principle*. London: Routledge and Kegan Paul. 1972.). Or call The Police (The Police. "Synchronicity I" and "Synchronicity II." *Synchronicity*. Record album. A & M. 1990.).

[21] When we first heard our teacher say "Macbeth" and "Macduff," the class thought we were going to be taken on a field trip to MacDonald's.

[22] Perhaps that should be "infamous," but we had not yet learned the difference between the two words until subsequently watching the Orion Pictures feature film, *The Three Amigos* (1986), in which Martin Short (as Ned Nederlander) explains that "infamous" means "*more* than famous."

[23] Produced largely on stage and using props leftover from previously filmed studio Westerns, some purportedly made of painted cardboard—a true challenge for media, technology, and imagination.

[24] I suspect that was Eagleton's point in commanding our undivided attention.

[25] I suspect that was the teacher's point in commanding our undivided attention.

questionable intention of my teacher who sat me in the rear of the room. From that bloodfire grew the threat that would tear open all the still-scarred wounds that my new teenage conformation and charm had marginally concealed.

First, Orson Welles had a big head, and so did I. Second, Macbeth had three doom-forecasting witches called *evil sisters*, and so did I.

It was clear to me that the three shadowed and cloaked figures standing atop a granite (cardboard) ledge with bifurcate twigs high above their heads had molded from the Scottish peat of a Hollywood set a heavy-headed voodoo model of me, and that that clay construction would collapse—no, be decapitated—in defeat by the end of the film. This was my life, played before all to see—not just my classmates but also people worldwide who spoke the twenty-three languages into which the film had been translated up to that time.[26] This was the impersonal manifestation of my own three elder and female siblings and their constant reminders of my Orwellian head and consequent impending doom. My fate was not just toil and trouble, but double-double so[27]—boiled and baked in a cauldron of unfailing anguish, a boiling hell-broth of

> Fillet of a fenny snake,...
> Eye of newt and toe of frog,
> Wool of bat and tongue of dog,
> Adder's fork and blind-worm's sting,
> Lizard's leg and owlet's wing...[28]

And lest we forget (and I never can) that *Macbeth* is all about *heads*—heads of kingdoms, heads of armies, even the head of Macbeth himself at the play's end, severed from its disproportionate (in Welles' case as in my

---

[26] This includes the one-time precocious blonde girl who played Venus, who I am certain spoke, kissed, and tickled in one of those languages which I had hoped would be French.

[27] Which, being inclined to math and sciences at that time, I inferred as a simple mathematical progression as Fate = 2 x 2 x (Toil + Trouble), as opposed to the much more serious exponential progression, Fate = $([\text{Toil} + \text{Trouble}]^2)^2$. Judging from ensuing events, however, I may have been wrong.

[28] IV.i.12-17. Little did Shakespeare, or the witches, whichever was real, realize that this is my great-grandmother's gumbo recipe. For those of you who intend to make some of your own, you can substitute a whole lizard for the two ingredients "lizard leg" and "blind-worm sting" because blindworms have no legs. To emulate the "sting" of the blindworm, a little Tabasco sauce will do. However, under all circumstances do not suffer the omission that Shakespeare has made in the *Macbeth* manuscript: First, you make the roux.

own) body and displayed for all conquering soldiers to gawk upon. But even more harrowing for the head-conscious student is the Armed (Armored) Head of Act IV, a bodiless apparition conjured by the sisters that speaks a mortal warning with a prophetic couplet[29] of lines: "Macbeth! Macbeth! Macbeth! beware Macduff; / Beware the thane of Fife. Dismiss me. Enough."[30]

And so, after suffering the corporal pyre of watching that film, will all my classmates mocking me through the backs of their heads, I came to an understanding—a *capitulation*, to use a big word with the same Latin etymology as "head." Just as Macduff revolts for Macbeth's head, so the world is revolted by mine; revolt in both cases illustrated by three evil sisters. It would be thus throughout my life if not for the rise of a being that would finally come to justify my forceps-bending birth.

Sophisticate Titles[31]

Thick, heavy, oil-based paint slathered over huge engines and machinery held down by humongous nuts and bolts. The deafening hum and heavy vibration of hot machinery in constant motion. Smudges of oil and mud at every corner, on every tally book, at every control panel, in every steel-toed shoe print. And everywhere the sweet odor of rust and diesel.

When I was in second grade, I got to visit my dad's rig. Dad was the tool pusher, and the "rig" was a drilling barge, with three floors of heavy machinery rising from the barge floor.[32] Atop (and through) the floors on one end of the barge was the drilling platform with a huge, steel drilling

---

[29] And that's all the head gets to say. So typical of theater and life that the more corpus-centered actors get the most lines.

[30] A subsequent apparition—a human child—tells Macbeth to "be lion-mettled, proud" for, as explained previously in this essay, humans lack the fortitude to point out illogical conclusions to lions (IV.i.87-8, 106). (Alas, another mew.)

[31] "Sophisticate Titles" was the polite, unrevealing phrase of the sign on the temporary wall beyond the comic book displays at my hometown drugstore. Beyond that wall were shelves of magazines that only adults were allowed to view and to purchase. On one Saturday visit, and with one peek while no one was looking, the phrase was no longer unrevealing to this mind. Behind such a wall is where Terry Eagleton might have conducted research on Mills and Boon novels (particularly the Blaze, Desire, and Spice imprints) for *Literary Theory*.

[32] The guy or gal in charge on site of a drilling operation.

derrick that in height was probably twice the length of the barge.[33] Aside the derrick on the other half of the barge and on the top floor were the bunkhouses, kitchen, and tool pusher's office where I hung out.

This was the early 1970s, when drilling and exploration off and around the Gulf Coast was at its peak. And just starting a well, or during down time between wells, there was little risk for bringing a kid aboard, I suppose. In any event, and although only under supervision did I visit the drilling deck and rooms of heavily painted machinery, I was generally relegated to the office, where I played with things like exotic drilling supplies, manual adding machines, slide rulers, circular graph paper, and unsmudged tally books. I learned sharp-edged, polysyllabic terms from pictures and drawings in catalogs and technical reports produced by companies like Hughes, Reed, names made by acronyms and portmanteaux ending in "CO," and the macho-sounding Baroid and Dresser Magcobar. The office was nearby the kitchen, where I discovered breakfast sausage, T-bone steaks, and other manly and meaty staples of the roustabout's and roughneck's diet that I had never tasted before either at home or at the elementary school cafeteria. This was a world away from the other—an advanced (although filthy and smelly) technological (and well-fed) playhouse of a world.

Soon, I had played with and tasted (from the office and kitchen, respectively) everything in that water-locked microcosm and got bored. In response, my dad called an incoming crewman to stop at his local supermarket to pick up some magazines and comic books. After several days of anticipation, the crewman eventually reached the office with a large bag of magazines, mostly sports and sophisticate titles, together with random comic books. Although I was permitted to view the sports magazines, of course I was not allowed to see the more sophisticate titles. However, I remember glimpsing the cover of a *Sports Illustrated* magazine with Joe Namath sporting a greasy-looking Fu Manchu-style mustache[34] and a *Playboy* issue not with a photograph of a woman on the cover but rather with a bunny-bustiere-bowtie cartoon.[35]

---

[33] My parents have told me that "Derrick" was their first idea for my given name. I suspect that, because the sight of my newborn form reminded them less of a derrick and more of a salt dome into which a derrick would drill, they abandoned the idea.

[34] This was a time when Joe only played a few games per year due to injuries. He was probably on the cover of this issue because around that time he had opened a series of bachelor bars in several major cities that hosted NFL teams, much to the dismay of Commissioner Pete Rozelle. When Rozelle demanded Namath's divestiture from the New York City operation, Namath, with a tearful television

The anticipation of getting new coming books outside of my control was a significant source of anxiety for me because the crewman knew nothing of my exclusive taste in comics. I was strictly a Marvel Comics boy, and I tended to shy away from the mainliners, like Spiderman and the Fantastic Four because those storylines were just too long for me and my meticulous attention span. I was attracted to the rarer headliners—hero and villain alike—characters like the gangly mythological dragon Fin Fang Foom,[36] the technologically advanced Titanium Man and Cobalt Man,[37] and the one-off, single-issue stories from *Tales of Suspense*, *Tales to Astonish*, and other variety series.

The bag of comics I was given was ultimately a mixed one, which was better than I predicted. Among this impromptu stash collected onshore without my input but for which I was obliged to show appreciation to a

---

broadcast, announced his premature retirement. Unfortunately for Rozelle, Namath was professional football's most popular and marketable player. A few months later, Namath returned to training camp and continued his injury-prone career in football, a game Terry Eagleton apparently does not watch and believes should be abolished (Eagleton, Terry. "Football: A Dear Friend to Capitalism." *The Guardian*. 15 June 2010. Web. 15 November 2012.) Of course, he was talking about *association* football (that's *soccer* for you Brits), but his sentiment, I'm certain, would be the same. Perhaps he, too, was named to school sports teams as equipment. Or perhaps he was relegated to kicker. But wait: Aren't they all kickers in soccer? Well, that explains that.

[35] This was a time when there was absolutely no reason to have a bunny-bustiere-bowtie cartoon on the cover rather than the anticipated alternative.

[36]Foom emigrated from a peaceful planet (Don't all dragons?) that suddenly set about to conquer other civilizations. Of course, he ended up trying to conquer ours because which other planet would pay 12 cents to read about their own invasion by a comic book character? Foom looks like a traditional Chinese dragon, and his base of operations is in fact in China. Seems to me the slow, economic conquering of civilization in the twenty-first century is much less messy than Foom's fiery, tail-swinging exploits of the twentieth, but I guess that would be hard to illustrate.

[37] These two characters (together with a group of superheroes from DC Comics called the Metal Men, which I soon stopped collecting because of the silly illustrations and storylines, like, for instance, the red, thermometer-colored Mercury Man dripping down the side of a laboratory where evil scientists [a concept that offends me professionally] were creating super magnets to destroy the obviously magnetic superheroes) led to my interest in the periodic table of the elements, the 100-plus names and symbols of which I could recite from memory in elementary school. Of course, and much to my dismay, no one at home or school was impressed because they expected that kind of spectacular feat from a child with such apparent cranial volume.

grown man with no taste in comics was a couple of *Archie* Comics[38], an *Iron Man*[39], and a *Sub-Mariner*. It was *Sub Mariner* #49,[40] and, much to my surprise, it immediately became my favorite mag.

I was attracted to that issue for then inexplicable reasons. Was it because I was offshore and the superhero represented a similar lifestyle surrounded by water? Was it Sub-Mariner's little ankle wings which reminded me so much of Greek mythology? Was it his hatred of humankind for polluting oceans (in some issues by drilling for oil) and destroying his home, Atlantis?[41] Or had I become over-sensitized by the multitude of calendars and other pinups taped and tacked to the filthy paneled walls of the men's bunkhouses depicting nude women feigning modesty while posing behind large, strategically placed wrenches and drilling bits, causing me to coincidentally become sensitized to his green, scaly, proto-Speedo bikini bottom?

Then it hit me like the subterranean screech of a diamond bit hitting a plate of iron ferrite.[42] I wasn't attracted to the Sub-Mariner at all. What kind of defeatist attraction would a young boy have toward a superhero who lay prostrate and incapacitated in the cover art, being dragged unconsciously—not even putting up a fight—by robotic minions under direct orders of the issue's primary supervillain? No, it wasn't poor ol'

---

[38] Which was way below my reading and interest level, although, because it was meat and I was on such a meat-eating drilling rig, the thought of Jughead's limburger cheeseburgers was a hungering interest to me, as was Veronica, who I thought was really hot (nearly a sophisticate title herself), even though that wasn't really a catch phrase for second-graders in those days.

[39] This one irritated me more than the *Archie* comic books because the pages were skewed when it was cut at the printer, and I needed perfect symmetry in both my comics and my life.

[40] *Sub-Mariner* #49 appeared in May 1972, which means that by the time I received this issue summer vacation had started and I was not a truant after all (Lee, Stan., Ed. "The Dream Stone." *Sub-Mariner*, 1.49 (1972): 1-21. Print.). Coincidentally, at the end of this issue, Subby ends up on the foggy, moonlit streets of the French Quarter of New Orleans embracing a brunette woman who looks remarkably like Veronica of *Archie* comics.

[41] Although Sub-Mariner no longer has his own comic book title and therefore reports of his antagonism of humankind are few, far between, and relegated to backstories of other superheroes who actually have their own titles, I wonder how he would feel today about British Petroleum. Somebody call the Marvel editors.

[42] A scenario borrowed from the asteroid drilling episode in the Touchtone Pictures feature film *Armageddon* (1998).

Subby at all. Rather, it was Subby's macrocephalic captor—the menace of Modok![43]

If This be Modok[44]

Now, Modok wasn't born a macro.[45] Like many superheroes and supervillains,[46] his powers were induced. But he wasn't bombarded with gamma rays (like the Fantastic Four), exposed to a nuclear blast (like the Hulk), or bitten by a radioactive spider (like Spiderman). Instead, Modok was originally George Tarleton, a mild-mannered scientist[47] working for Advanced Idea Mechanics, better known as AIM. AIM had recently created the Cosmic Cube, a device designed to further their ultimate goal of world conquest, but, due to its reality-reshaping, trans-dimensional, portal-opening powers, a device that they couldn't quite understand; they

---

[43] Modok, which has been called "one of the greatest visuals ever" by then Editor-in-Chief of Marvel Comics Joe Quesada (Tarleton, George. "Here's Why Joe Quesada Should Just Go Ahead and Sell Me Rights to Modok." *Journal of Modok Studies* 1.2 (2004). Print.), was created in 1967 by Marvel editor Stan Lee, illustrator Jack Kirby, and embellisher Joe Sinnott. "Modok" is an acronym for "Mental Organism Designed Only for Killing," although he was originally intended to be "Modoc" (Designed Only for Computing).

[44] This subtitle, in the stark, present subjunctive case common to the Silver Age of Marvel comics, is also the title of *Tales of Suspense* #94, the comic book debut of Modok. (Lee, Stan and Jack Kirby. "If This Be Modok!" *Tales of Suspense* 1.94 (1967): 1-10. Print.)

[45] Although Modok gloriously appears first in *Tales of Suspense* #94 (from 1967), the story of his origin is first told three years later in *Captain America* #133 (Lee, Stan and Gene Colan. "Madness in the Slums." *Captain America* 1.133 (1970): 1-18. Print.), most elegantly told in the non-graphic novel *And Call My Killer Modok*, by William Rotsler (New York: Pocket Books, 1973. Print.), and somewhat less elegantly revisited in "The History of Modok, Part One" (*Journal of Modok Studies* 1.1 (2003): 2-5. Print.)

[46] All except the X-Men, who are declared mutants with special gifts conferred at birth. Genetically, their gifts (or, better, mutant phenotypes) are congenital. They are mutants like all of us. Current genetic understanding suggests that we are all born with about 60 mutations different than our parents, products of spontaneous changes in their germlines (Conrad, Donald. "Variation in genome-wide mutation rates within and between human families." *Nature Genetics* 43 (2011): 712-714.). Of course, if a germline was exposed to radiation like many Marvel superheroes have been, then maybe it would be something like 6000 times 60 mutations. But, realistically speaking, those exposed wouldn't likely have a functioning germline at that point (Somebody call Sue Richards.). So who's counting?

[47] There are so very few of us.

needed an advanced mind—a controllable mind and one capable of human communication—to understand and wield the Cube according to their world-dominating wishes. And George Tarleton was the path of least resistance—the scapegoat, the guinea pig, the *Drosophila*[48] for their next creation.

Tarleton, placed into the "alteration chamber," was within seconds transformed[49] into "…Modoc…the most powerful brain alive!" as one of his AIM captors explained.[50] Having the most powerful brain alive is a cool thing, Tarleton's ego must have felt,[51] but that brain was contained in a cranium of circumference that rivaled a small moon of Jupiter. He had become a hideous being with an enormous head and a disproportionately flailing body that was no longer the focus of either his motor skills or his being. Fortunately, AIM had invented a really cool hover chair[52] for the flailing Tarleton that just happened to be the perfect size and shape to support the massive and pudgy head of the victim of a process they obviously didn't know how to control. The originally shiny silver chair (later a school-bus yellow) held the gargantuan head in place while the rest of the body fit into the base of the chair. Across Modoc's forehead was a headband[53] that, fitting perfectly around his long-banged, bowl-cut coiffure, focused his concentration into powerful psionic waves[54] with

---

[48] *Drosophila melanogaster* is the scientific name of the common fruit fly, which since the 1910s has been a routine model organism for studying the effects and transmission of mutations in laboratories and teaching classrooms.

[49] Though a process that has been called "advanced mutagenics" and not explained further. However, forty years later, we are clued. In *Marvel Adventures: The Avengers* #9 (2007), Modok and AIM have redesigned the alteration chamber into a "cellular restructurer" which operates, according to the commentary of the AIM operator, as follows: "Initiate reformation sequence...DNA chains unlocking,... commence cerebral expansion,..." and *voila!* (Pannicia, Mark, Ed. "A Not-So-Beautiful Mind." *Marvel Adventures: The Avengers* 1.9 (2007): 6. Print.).

[50] *Captain America* 1.133 (1970): 4.

[51] And it was now a very large ego. Even his superego was supersized.

[52] Actually, from Page 4 (*Tales of Suspense* #94), it is a "magnetic-powered mobile chair," far more agile than the naturally occurring legs of the infinitely more athletic Captain America. But it also hovers.

[53] After a whole decade, Modok's fashion statement became vogue in popular culture, as evidenced by musical acts of the late 70s and early 80s, such as Loverboy and later Olivia Newton-John—although neither generated much psionic influence.

[54] Actually, on Page 5 (*Tales of Suspense* #94) Modok describes it as his "invincible mind beam... the greatest single weapon ever unleashed!!" But he also has a stun-shock beam and a heat beam.

which he could directionally blast friend or foe to influence or to ruin. His brain emerged with superhuman calculating ability, and he can recall any fact he has ever known. In addition, with the rise of such technology in our own universe, he can translate himself into raw data and travel through email.[55]

It wasn't long, however, before Modoc (Mental Organism Designed Only for Computing) became Mental Organism Designed Only for Killing (Modok). Still in the alteration chamber, the former Tarleton had come to realize that his power was now greater than that of his captors, whom he promptly destroyed[56]—probably only after they dressed him in that perfectly-fitted hover chair. In just a moment, the pent-up anger and discouragement over his inflated head transformed a mild-mannered mathematical genius into a calculating destroyer of men.

Modok drifts through the Marvel Comics universe over the next half century, bothering many superheroes. Like each and every one of their other superfoes, Modok never quite accomplishes effective victory—much less complete victory—over these heroes, only tearing a shard of red, molded sheet metal here (from Iron Man) or a tatter of blue stretchy fabric there (from Captain America). The narrative excuse for this unreasonable, five-decade game is this: Because his mutagenized brain has been given over to computing, those neural centers for creativity are no longer fully functional.[57] He is, if you will, "too calculating" to envision an indefensible, unpredictable end to any superhero. This fate is prophesized not long after his origin (only 27 frames), when, on the final page of his first comic book appearance we see a destroyed Modok, lying supine on the ground amid the blasted shrapnel of his hover-suit, his feeble limbs contorted and lifeless, his large, agonized face beneath his mussed coiffure, bemoaning death "just like any helpless being."

*For...once I possessed the power...to change the world!! Therefore,... I must perish...in a manner...befitting.,.**the master supreme**!!*[58]

---

[55] As of 2013, however, Modok cannot tweet himself.

[56] Take heed witches, teachers, coaches, neighbors, moviegoers, kid-contemporaries, piñata-poppers, Ahabs, fifteenth-century Portuguese sailors, all.

[57] "History of a Mentalist." In Cowsill, A. (ed.) *The Classic Marvel Figurine Collection*. London: Eaglemoss Publications, Ltd. 2011. P. 3. (among other places in Marvel literature).

[58] This is his requiem on the final page of *Tales of Suspense* #94 (P. 10). The ellipses are not those of the author to indicate missing information but rather because Modok is gasping for air as he speaks these words. This sympathetically illustrated panel is recapitulated at Modok's death in *Captain America* #313, when Captain America is moved to comment that Modok "didn't deserve to die like

If I be Modok, then what is Literature?

At that point, energy beams are illustrated in red and yellow, overlaid with some onomatopoeia spelled with letters that each earn lots of points in *Scrabble*, indicating that the inevitably vengeful Modok blows something up and then perishes. Or so we are led to believe. Modok is resurrected from that fatal 1967 event, which, as it turns out, was only *near* fatal. Dozens of times over the ensuing decades, his perishings are only near fatal,[59] with alternating control of and control by the agents of AIM.[60] Despite the occasional sympathy of Captain America, which is always expressed only after his complicity in the near fatality is over, it is as if no Marvel story writer would help Modok to fully comprehend defeat or better yet win some sort of subsequent victory—to take the super calculable to satisfactory completion.

Though there is good and continuous evidence that computation and creativity utilize opposing hemispheres of the human brain (left and right, respectively) to different extents,[61] it is clear that both hemispheres are utilized synchronously. It is therefore difficult to comprehend how an "alteration chamber" bombarding George Tarleton with some sort of sprayed energy field would have enhanced one hemisphere and not the other.[62] Perhaps the energy ray source on the right side of the chamber was blocked by cobwebs, duct tape, or splattered remains of a previous test organism. Perhaps Tarleton had rolled onto his right side as he struggled for freedom on the floor of the chamber, leaving the left side of his cranium to be more directly exposed.

---

this." (Carlin, Mike, Ed. "Mission: Murder Modok." *Captain America* 1.313 (1986): 22. Print.)

[59] The one true fatal episode that appears in *Captain America* #313 (1986) has Modok bitten by a poisonous snake-man of the Serpent Society, who then returns the lifeless Modok body to AIM for use as a supercomputer. In their finite wisdom, AIM creates a new Comic Cube and a sudden need to resurrect Modok in *Iron Man* #205 (Gruenwald, Mark, Ed. "Modok is Dead..." *Iron Man* 1.205 (1986): 1-20). So, to invoke Mark Twain (an author, by the way, like Orwell, also not mentioned in Eagleton's *Literary Theory*, perhaps because to Eagleton his fiction is a bit comic-bookish), the rumors of Modok's death were, like his big head, greatly exaggerated. Modok actually quotes this quote in the issue, but, of course, he doesn't say "like [my] big head."

[60] For a comprehensive narrative sequence of Modok's near fatalities, see: "Modok." *Marvel Universe Wiki*. Marvel. Web. 15 November 2012.

[61] For a recent study, see: Kounios, John, et al. "The Origins of Insight in Resting-State Brain Activity." *Neuropsychologia* 46.1 (2008): 281-291.

[62] Someone call Dodge. Modok's got a *hemi*.

Whatever the logic for the disposition of Modok, the often overlooked fact remains that only moments after the mutagenesis—in the spontaneous "Aha!" moment that changed him from *Modoc* to *Modok*—he *created* a new direction, he commanded a new, unprescribed fate. His "designed only" life became a "designer" life on its first day:

Yes. Once I was a mere experiment for the scientists of AIM, and I in turn became their ***master****. My creators did their job too well.*[63]

"Aha!" moments, the hallmarks of creative problem solving, are also landmarks of our early development as sentient beings, as well as during adulthood. Such thoughts are those coffee-spilling moments that earn praise and occasional choking from our daydreaming and disinterested colleagues during committee meetings. From a genetic standpoint, you could say that "Aha!" is part of our inherited *design*. Even in deep computational reverie, the human mind is powerless to avoid even intermittent "Aha's."[64]

Then, I pose to you, my friends, this: Are we not *all* mental organisms? Are we not *all* designed for creativity? Are we not, then, Modocs ourselves? And if we are indeed born Modoc (Designed Only for Creating), wouldn't our cultural literature include the *Tales of Suspense* in which we were first worded?

And then I shall become ***literary theorist supreme****! But, apparently not to all those who are what I can never be again share this sentiment. They must know Modok's price for resistance!*

There are some, however, who would agree with Terry Eagleton in *Literary Theory* that "[comics] ... are ... not generally regarded as literature, and certainly not Literature."[65] Presuming both Eagleton and the typesetter at University of Minnesota Press used healthy, fully-

---

[63] Paraphrased from *Tales of Suspense* #94 (P.4). As used here, this is now Modok channeling though the computer with which this essay is being written. The italics are his. Despite the fact that insertable callout shapes are available in this version of Microsoft Word, they cannot point to an appearance of him due to copyright restrictions and are thus not used to denote his speaking. *An infinitesimally small subroutine in a piteous algorithm cannot match the immense calculability of* ***Modok****.* That's right, Modok, but Disney, which has owned your copyright since 2009, sure can. I'm certain you immense calculability can figure out legal defense fees for that. And can you please leave the footnotes to me? So, reader, you can consider these upcoming italicized Modok-speak lines as redacted frames of an unprintable Modok comic book appearance.

[64] Think of Albert Einstein finally placing the exponent on his otherwise dull, insufficient, and old-fashioned Newtonian-sounding equation, $E = mc$.

[65] P.2. The ellipses are not those of the author gasping for air as he writes these words—as far as can be told.

functioning keyboards[66] during production, capitalization of the second instance of "literature" was intentional. Hiding behind the "not generally" modification will only go so far toward protecting the author from criticism by us Modocs, but the reinforcing "certainly not" is too difficult to accept, particularly because it is not farther than three paragraphs into his book-long diatribe.[67] Is this condemnation society's criticism? Or Eagleton's? Or Eagleton *directing* society's criticism?

It pains Modok to hurt one with such a great cranium, but his verbosity has left Modok ***no choice****.*[68]

The new peaceful, creative Modoc society, however, has several points with which to dismiss such deprecation. First, what "generally regarded" opinion can be assimilated from one who offers one book of criticism (*Literary Theory*) then writes a sequel criticizing the previous (*After Theory*),[69] and enjoys royalties from both. Second, those who have read *any* of Eagleton's writings know that he can't go very far into a paragraph without critically insulting someone or condemning some part of society,[70] so why should Modocs expect to be treated differently? Sure he's Marxist, but, really, can he not enjoy sipping a cup of Twinings or Tetley without condemning the politics of whether the handle of his tea cup is facing toward the left or toward the right?

Modok hears Eagleton's piteous tea-sippings. His great cranium is as glassible as the miniscule cup he holds with his left hand. Modok melds with his mind[71] and learns that this one will not relinquish his role and serve as Modok's ***subordinate****.*

Third, although he suggests sympathetically that works like comic books

..[have been] hermetically sealed from history, subjected to a sterile critical formalism, piously swaddled with eternal verities and used to

---

[66] Those unhampered by polymerization of desiccated orange juice or sweetened coffee spilled over one or the other Caps Lock key.

[67] One of his many.

[68] This marvelously appropriate line is approximated from a win quote that derives from the *Ultimate Marvel vs. Capcom 3* video game, in which it appears in the first person. ("Win Quote vs. Viewtiful Joe." *Fight-a-Base.* Web. 15 November 2012. <fenixware.net/fab/fab_quoteMain.asp?id=19710>.)

[69] If I offered the title *Enough Theory*, then that would be as much a pun as the point.

[70] Even in one *Wikipedia* article written by other hands ("Terry Eagleton." *Wikipedia.* 4 January 2013. Wikimedia Foundation. Web. 6 January 2013), he insults five writers in as many paragraphs. And that's on a slow day.

[71] A trick Modok learned from Spock.

confirm prejudices which any moderately enlightened student can perceive as objectionable.[72]

This sympathy comes only after *ninety thousand words* and—worse—accompanied by no illustrations. There in the conclusion to *Theory*, we must ourselves conclude that Eagleton has finally realized the relevance of our literary culture—that the swaddled academic instruction hermetically sealing our literature from history has, in fact, served as our alteration chamber, creating mad mental organisms designed only for conformity. We are indeed Modocs, borne of a life in four colors stamped onto newsprint, and though he or history or society or the academic traditions of English literature may attempt to condemn us, after spilling many words he realizes that he can no longer risk being part of that contempt. Has he joined us? No. Perhaps, though, he was visited by witches conjuring apparitions, one of which was a bodiless head bidding him beware. Or, perhaps:

Look out, Terry, 'cause Modok's using technology![73]

at which point, in the anxiety fibers of his imagination center deep in Eagleton's underutilized right hemisphere,[74] energy beams of a psionic blast appeared in red and yellow, sounding some onomatopoeia spelled with letters that each earn lots of points in *Scrabble*.

After that hallucination, Eagleton leaves us in *Literary Theory* with a lion-metaphor, an allusion perhaps to the Macbethean apparition.[75] In true revolutionary spirit,

---

[72] *Literary Theory*, 216-7.

[73] "Search and Destroy."

[74] And it must be the right one at that moment, for, after spilling ninety-thousand words, surely there could have been no previous *Aha!* from the left hemisphere. As Eagleton's mind is "left-leaning," perhaps, too his brain is left-hemisphering.

[75] But not an allusion to George Orwell, whom, as we know from Note 19, is largely disregarded in *Literary Theory*. However, Orwell was not unacquainted with the occasional lion-metaphor, as shown by his essay, *The Lion and the Unicorn* (1941), in which the lion (yet another brinded mew) is a symbol of revolution, just like Eagleton's 42 years later. *The Lion and the Unicorn* begins with the famous (infamous? See note 22) line, "As I write, highly civilized human beings are flying overhead, trying to kill me." I predict that Eagleton's projected next essay, *Enough Theory* (see note 69) will begin, "As I criticize, highly designed mental organisms are hovering overhead, trying to psi-blast me." In addition to lion-metaphors of revolution, Orwell anticipated Eagleton in another important way: The *thoughtcrime* concept of his novel, *Nineteen Eighty-Four* (f.p. 1949) was devised to try to encourage the six-year-old Eagleton (b. 1943) to keep his future plethora of ninety-thousand-word opinions to himself.

We know that the lion is stronger than the lion-tamer, and so does the lion-tamer. The problem is that the lion does not know it. It is not out of the question that the death of literature may help to awaken the lion.[76]

Politics aside, and judging from the relative sales of comic books versus Eagleton books, we Modocs would say that the lion has been awake for a long period of time. And *this* lion is one, big, honking cat.

Consequences of the Design

And Modok himself has remained awake. In numerous incarnations, which include a female Modam (Designed for Aggressive Maneuvers) and a canine Modog (Designed Only for Genocide)[77], the idea of the designed mental organism has permeated the Marvel Universe,[78] as well as popular culture.[79] These days, macrocephaly is cool.

And unless you've kept up-to-date with recent discoveries in the field of childhood development (and who hasn't), you will have missed the arguably most important discovery in the history of megacephaly study. It's a discovery that vindicates my life. A recent report in the medical journal *Pediatrics*[80]shows data that actually tells something positive about children with big heads:

---

[76] *Literary Theory*, 217.

[77] Not to be confused with Moondog, the late New York street musician and contrapuntal composer who was a popular contemporary of Modok during the days of the first Comic Cube, releasing several albums of "found" music with instruments of his own invention.

[78] In *Marvel Adventures: The Avengers #9* (2007), the new cellular restructurer converts the Avengers of that year—Captain America, Storm, the Hulk, Spider-Man. Giant-Girl, Iron Man, and Wolverine—into the macrocephalics Capdoc, Stormdoc, Spidoc, Hulkdoc, Gi-doc, Irondoc, and Woldoc. This time, the victims were placed into hover suits first, making the process much less difficult to imagine than that of *Tales of Suspense* #94 (1967). At this point, Modok renames himself with his original acronym—Modoc—this time meaning "Designed Only for Conquest." For a partial list of designed mental organisms, see: "Modok." *Wikipedia.* 15 November 2012. Wikimedia Foundation. Web. 15 November 2012. And lest we forget, see also Season 1, Episode 15 of *The Super Hero Squad Show*, entitled "Mental Organism Designed Only for Kisses" (2009), available on DVD and YouTube.

[79] Simply Google for images of "modok" and see what happens.

[80] Gale, Catharine et al. "The influence of head growth in fetal life, infancy, and childhood on intelligence at the ages of 4 and 8 years." *Pediatrics* 118 (2006): 1486-1492. Print.

Brain growth in early life may be important in determining not only the level of peak cognitive function attained but also whether such function is preserved in old age. Older people with a larger head circumference tend to perform better on tests of cognitive function and may have reduced risks of cognitive decline and of Alzheimer disease.[81]

Despite long-lasting emotional (and concussional) traumas, the other long-lasting effects of big-headedness in kids—according to the journal *Pediatrics*—are increased intelligence and decreased incidence of Alzheimer disease later in life.

With my painful childhood vindicated, and perhaps that of some of you more silent Modocs out there as well, I conclude. For all my acorns are now buried.

And I am lion-mettled.[82]

---

[81] Gale, 1491.

[82] And perhaps, too, a bit brinded, but in a good way. Mew.

CHAPTER TWO

THE AGGREGATE SPECTATOR: 125 YEARS OF SIGHTS, SOUNDS AND MOVING PICTURES

DEL JACOBS

Let's start with the aggregate spectator—the moviegoer. This phenomenon is of interest to a number of disciplines. The historian is concerned with "what" we watch. The sociologist with "how" we watch, including with whom and in what environment. The psychologist with 'why' we watch, for there is a therapeutic function in all media. These all lead to the 'result' of watching, obviously cross-disciplinary, and that gives us the aggregate spectator today—us, you and me.

That first generation film audience experienced the sheer novelty and power of watching pictures that move. From 1895 through the 1920s they witnessed the birth, development and consolidation of an art form and industry. Some films were tinted or hand-colored. Some experimented with screen or image size. Some were even synchronized with audio.

Indeed, most in that first wave consumed movies that were silent, or accompanied by piano players, organists or other musicians, even symphony orchestras. Complete films could be 30 seconds in length, as were the first Lumière and Edison works, or an evening's worth of entertainment such as Griffith's feature-length *The Birth of a Nation* (1915). Many of the genres celebrated today were established and experienced early codifications during this period: the historical and contemporary drama, romance, western, crime, animation and comedy.

It is around this time that the growing influence of film initiated a modern culture of celebrity. This was helped by the publication of fan magazines like *Photoplay*, with its first edition in 1911. By the time sound pictures become an industry-wide reality, the first generation of spectator was giving way to the second generation. The years from the late 1920's through about 1950 comprise Hollywood's Golden Era. The studio system was firmly established and its factory-like output to a mostly homogeneous audience set a worldwide standard for quality, making American movies the desired import in most foreign countries.

Movies came to grips with their role as a moral arbiter that dodged or encrypted certain volatile issues and representations, especially after the Production Code of 1934 was established. The young viewer who saw Lon Chaney, Sr., in *Phantom of the Opera* in 1925 or Fredric March in *Dr. Jekyll and Mr. Hyde* in 1932 would find these characterizations tamed somewhat by Claude Rains (*Phantom of the Opera*, 1943) and Spencer Tracy (*Dr. Jekyll and Mr. Hyde*, 1941) in the post-Code days.

Hollywood films provided both an illumination of social issues as well as a forum for escape during the Great Depression. Most studio movies produced during World War II contain prominent home-front themes as they bolster morale for the war effort. During this time attendance at American movie houses reached its all-time high, with about 95 million tickets often sold weekly during the mid-1940s.

After 1945 and the acceleration of postwar institutions such as suburbia and television, a third generation of film audiences emerged from the baby boom. But let's not leave this second generation of spectators just yet.

Second generation audiences spoke wondrously of the year 1939, considered to be the most "golden" of the golden era. The production output was certainly vigorous that year, with 476 U.S. pictures released. Contrast this with 143 films released in 1971, one of the least productive years before production titles increased due in large part to emerging video and cable markets of the 1970s. Two movies from 1939—both from the same studio (Metro-Goldwyn-Mayer)—crystallize the moment-as-memory for many of that generation. *Gone with the Wind* is, adjusted for inflation, the most financially successful American film in terms of box office gross revenue. Its audience was pre-sold by the popularity of Margaret Mitchell's novel on which it is based as well as the cast and high-quality production values. It won eight Oscars and was reissued frequently before its eventual release to television and home video. If *Gone with the Wind* is about the mythic American past, then *The Wizard of Oz* is a counterpart that also gives us a world of fantasy complete with its own mythic underpinnings. *Oz* became a perennial favorite on television screens before *Gone with the Wind* was released for broadcast, and thus probably initiated many second and third generation audiences in the moving picture experience . . . and in the fear of flying monkeys!

Third generation spectators incorporate the phenomenon of television into their viewing experiences in the period from the early 1950s through the mid-1970s. In 1948, only 58 American cities had TV stations and they broadcast to 1.6 million sets. People watched TV communally—in a bar, a bowling alley, at the home of someone who could afford a receiver that was often more furniture than screen. But by 1954, there were 32 million homes with television and Hollywood knew it had a competitor. Attendance at theaters confirms this, with weekly ticket sales at about 47 million in 1957, half the wartime peak. As the decade ends, 90 percent of homes had television.

Hollywood reacted with fear, hostility and innovation towards the new medium. Movies themselves—like Elia Kazan's *A Face in the Crowd* (1957) starring Andy Griffith—would be outright critical of television in the course of their narratives. Jack Warner, in fact, banned TV sets from the living room sets of his movies for a period. The film industry also supported an ad campaign called "Movies Are Better than Ever" throughout the 1950s with a series of presentation gimmicks, some of which proved short-lived and others long-lasting. The nature of film spectacle was enhanced by widescreen, the curved screen of Cinerama, stereo and multitrack sound, 3-D movies, and more carnivalesque ploys such as Aromo-Rama, Perceptio!, Duo-Vision, Emergo, PsychoRama and others, right up through Sensurround in the 1970s. The films of William Castle were thoroughly enjoyable excursions into the ballyhoo of selling low-budget genre films of the time like *13 Ghosts*, *Macabre*, *The Tingler*, *Homicidal*, and *The House on Haunted Hill*.

While narrative often plays a secondary role to spectacle or sensation, the movie screens themselves in the early 1960s actually became smaller. The large, urban cinema palaces of the silent and early sound era suffered attendance drops during the 1960s and 1970s. Many were demolished, abandoned or retrofitted as multi-screen showcases. Some gave themselves over to programming for inner-city audiences since the core middle-class filmgoers prefer suburban outlets for movies that are closer to their homes. Drive-ins proliferated in the 1950s. More than 4,000 were operating by mid-decade, when they accounted for about one-quarter of the U.S. box office revenue. In June 1957, nearly 6,000 drive-in theaters were standing in America. One week that summer their attendance actually superseded that at indoor theaters for the first and only time.

By the early 1960s, multiplex cinemas began to dominate exhibition practices. Stan Durwood and American Multi-Cinema pioneer the multi-screen theaters, largely at shopping centers, at first. The film audience—once unified by choice and design at the single-screen theater—became fragmented into niches so that even families can choose different films on a night out, much like they selected channels on the television sets in their viewing rooms at home.

Besides the diversity in technologies and in the screening venues, there is the content of cinema that distinguishes itself for third generation spectators. American movie directors—at first influenced by the Italian Neorealists and the French New Wave—began to make more personal films under new banners of expression in the 1950s and 1960s. As in other periods of history, the popular arts during this time experienced a reevaluation of their boundaries of expression. This included forays by Elvis Presley and Little Richard in popular music, Hugh Hefner in the magazine publishing world, Jack Kerouac in modern literature, Jackson Pollack and Mark Rothko in the art world and Andy Warhol in popular fashion.

The sociohistorical scene was ripe with tensions that fuel inspiration for filmmakers in this generation—the Cold War, McCarthyism, civil rights, Vietnam, feminism, generational conflict, the drug culture, political assassinations, identity politics, environmentalism and more. Movies may have become more diverse, but annual attendance diminished to an all-time low during this period, with 1971 the nadir of a 25-year economic decline after the peak year of 1946. This is not to imply that engaging and ground-breaking mainstream motion pictures were not being made. The year 1967 produced *2001: A Space Odyssey, Bonnie and Clyde, The Graduate* and *Rosemary's Baby*; 1968 showcased *Midnight Cowboy, The Wild Bunch, Z* and *Night of the Living Dead*; 1969 presented *Easy Rider, Butch Cassidy and the Sundance Kid, Medium Cool* and *They Shoot Horses Don't They?*; and 1970 brought *M*A*S*H, Catch-22, Little Big Man* and *Five Easy Pieces*. Generally, films became more youth-oriented as the older "lost audience" preferred television.

What is the demarcation point between third and fourth generations? As with the junctures that came before, it is less easy to point to a specific moment in time than to a collection of events in roughly the same period. Dominating would be two technological occurrences—the convergence of cable TV systems, with special attention to Home Box Office (HBO) as a premium service in the mid-1970s, and the market penetration of the videocassette recorder (VCR) thereafter.

In the film content arena, however, there was the new importance of the "blockbuster" movie that could once again reach a mass audience, invite repeat viewings, engage watchers through some new parity of story and spectacle, create synergy with other consumer products and perhaps spin off sequel films that instigate further products. First, *Jaws* (Spielberg, 1975) and then more precisely *Star Wars* (Lucas, 1977) epitomized this strategy. Blockbuster movies became "experiences" and cultural phenomena.

If the third generation encountered the tension between the movie experience and the television experience as one of competing and coping, the fourth generation of audiences saw the boundaries between the two mediums of film and TV become both blurred and porous.

Video became—in the fourth generation—the primary means by which audiences experienced the moving image of narratives originally termed "film." Annual revenue from videocassette sales and rentals outdistanced film box office receipts for the first time in 1987. Competing formats like the videodisc had a brief run of success. People could proudly display their video collections. Since that time—when considering all then-current video-on-demand services, video games which incorporate narrative and spectator strategies, and home video as a secondary market for movies—more people "went out to get a movie" than "went out to see a movie" during any given week.

Yet, the big-screen experience is still—then and now—the launch-point for many films that become successful in their secondary market exposures. A screening room in a private home where current theatrical releases would be played was formerly the purvey of Hollywood movie moguls, members of the jet-set, or the occupant of the White House. But since the 1970s, recent feature films are delivered to homes wired for cable and subscribing to HBO, Showtime, and several other movie channels. The box in the corner that formerly delivered mostly G-rated fare for the entire family now brings films and comedy specials geared to more mature tastes.

Coupled with the videocassette—which was a consumer product that earned its market penetration as much with adult films (i.e., pornography) as with programming for other niches (exercise videos, sports programs, music videos)—consumers began managing their own presentations in their own private spaces to fit their own schedules. The affordability of home video camcorders unleashed a renewal of amateur or neophyte moviemakers who capture reality on their own and then instantly replay the images. Since the 1920s, amateur movies—mostly referred to as "home movies"—had been economically feasible for some consumers. The emergence of 8mm and Super 8mm film, and then videocassette tape, provided the impetus for many to "write with light" in the latter decades of the 20th century. Undoubtedly, many in the fourth wave witnessed their first moving picture experience as home movies, starring themselves.

Along with the blockbuster movies that became cultural events during this period (including *The Godfather* films, *Titanic*, and *The Sixth Sense*), there are reactionary forces. The rise of independent cinema in the 1980s and 1990s—continuing through to today—afforded a platform for low-budget films, formerly marginalized topics, and allowed maverick moviemakers to enter the game. Multiplex screens, cable channels, and video stores beckoned new producers who, while rarely reaching the blockbuster status, were able to gain notice and perhaps access to a larger professional film arena than ever before.

When not able to create spectacles due to budget constraints, theses indies often relied on quixotic narrative techniques making their movies a minor-league cultural experience. The least expensive and most innovative notion to employ is in editing, via the disruption of linear narrative flow. *Reservoir Dogs* (Tarantino, 1992), *Run Lola Run* (Tykwer, 1998) and *Go* (Liman, 1999) have given way to further structural ruptures in recent films such as *Memento* (Nolan, 2001) and *Irreversible* (Noe, 2002), right up to *Inception* (Nolan, 2010) as we get closer to the fifth generation of spectators.

Another low-budget technique that distanced independent film from blockbusters in the fourth generation is the very "look" of the movie. *The Blair Witch Project* (Sanchez, Myrick, 1999)—a pseudo-documentary—was purposefully shot in grainy 16mm and on videotape. Established filmmakers, such as Spike Lee in *Bamboozled* (1999) used digital video, and that technology slowly worked into the blockbuster mentality. Some big-budget ground-breakers employed this technique. *Star Wars: Attack of the Clones* (Lucas, 2002) was photographed and, in some cases, projected in digital video, and *Collateral* (Mann, 2004) alternatingly used a mixture of 35mm film and digital video.

The fourth and fifth generations overlap, but by the new millennium, we are grounded in the fifth. The moving image is ubiquitous and omnipresent for the fifth generation of spectators. For many of us, the difference between celluloid and digital video may be a moot point. Today, cinemas are rapidly converting to digital video projection of "files." Purists will continue to argue about the warmth of film stock versus the sterile sheen, graininess, or low contrast of video.

While the quality of image will matter in many cases, the ease with which images are created and called up is most significant for the fifth generation. An internet platform such as YouTube permits anyone to upload and share moving images and sound with anyone who has access to a computer screen. For every minute that passes in real time, 60 hours of video are uploaded to YouTube. That's 10 years of video every day. More video is uploaded to YouTube every month than has been broadcast by the three big TV networks in the past 60 years. Two technological delivery systems have revolutionized the way all generations today consume the moving image: the internet and the DVD (known as digital versatile disc, but more commonly digital video disc). At first, brief, flash-animation on website pop-ups, or short films, caught the fourth and fifth generations' attention. There is a constant process of reviewing and rejecting or accepting detours from the original quest for engagement or information. Soon, clips and programming of all variety loom. Feature-length movies can be streamed or downloaded from the web for uninterrupted, attentive viewing.

These options represent the breadth of moving images on computer screens—from a few seconds to feature-length—with a variety of other styles and dimensions persisting between them. A thorough exploration of the fifth generation of audiences for the moving image must begin, if not center upon, the home computer screen as a point of initiation and primary engagement. Since 1997, the DVD player began to reshape the way film is experienced. The movies were a natural data field for the DVD and that technology has changed not only the ways that movies are made, but the ways they are watched.

Two decades earlier, a movie on videocassette was about the size of a paperback book. Indeed, when Akira Morita of Sony Corporation instructed his engineers to devise what was to become Betamax tape, he requested it be in the relative size and shape of a book, the better to be collected and placed on the shelves of consumers' libraries. That shelf-space, today, is in the cloud or on a hard-drive for many of us. The film itself—in content terms at least—is identical to that same film projected in theaters. Nothing more, sometimes less. Three decades ago that version of a 35mm "film" would have been encircled on metal reels holding about 20 minutes each, with a feature encased in two steel canisters containing three reels each. The film weighed forty pounds or more and, in some theaters that were highly unionized, could only be carried to the projection booth and screened at the hands of a union projectionist. For many spectators, part of the mystery and magic of movies incorporated not only the anticipation of the movie-going event, but the fading of house lights in the cinema and the strike of the projector's arc lamp, then evidenced by the beam of light that floods out from the projection room portal, giving screen life to the celluloid images.

Today, the secrets of moviemaking are revealed ad nauseum on selected DVD audio tracks and supplements. A culture of cinephilia has demystified the movies. At the same time, the fortification of the "home theater" has turned middle-class consumers into varying levels of audio-video technophiles who orchestrate the presentation. So in many ways, the film storyteller has surrendered to the market force of narrative packaging in this war for the audience's attention. Is a viewer more, or less, empowered when she is able to literally stop the film and freeze the frame for whatever reason? Perhaps to answer the telephone, go to the bathroom, explain a plot point to a less involved viewer, or maybe even ponder the jitterless image of the picture-perfect frame in this pause mode, freezing and breaking into it in new ways?

The storyteller no longer controls the ball, as they say in sports, at least not in the way once done in the singular screening space of the cinema. The work of art/artist is now subject to manipulation as much as the spectator in the audience. Several aspects of moviemaking guided by the relationship of viewer to film are now mediated by television, the DVD, files and downloads and various viewing platforms. What will compel us to watch and how will we watch it all? As storytellers strive to continue to make frames worthy of watching, audiences have amped up their own strategies for maintaining or breaking them. Robert Ray's reflection on this postmodern experience is, "The art object no longer controls our reading of it." The fifth generation of spectators will have difficulty, it seems, viewing art for art's sake. But it confronts a mediated world with several opportunities for interaction with moving images that dwarf the choices of previous generations that were faced with cinema and television as the major means of spectatorship.

Indeed, the fifth and sixth generation of spectators—whether going to the movies, using the internet, or sitting before high definition video screens large and small—will be marked by some attributes shared by the aggregate listener which I will discuss after surveying those generations.

But Fifth Generation listeners have been weened on the Internet, which offers extraordinary opportunities for creativity, exploration, self-discovery and self-definition. While the individual ramifications are many and varied, I invite your own impressions and experiences to define yourselves as the fifth and sixth generation spectators and listeners. This is the "aggregate" of which we are one and all. There are attributes that are well worth marking. Just as technology changes the spectators' ways of experiencing the moving image, it also changes their means of responding to it. Over a decade ago Douglas Rushkoff coined the phrase "screenagers" to describe the first generation that grew up with the assumption that the images on a television screen were supposed to be manipulated, that they weren't just there for passive consumption. The fifth and sixth generations are carrying that logic to a new extreme: the screen is not just something you manipulate, but something you project your identity onto, a place to work through the story of your life as it unfolds.

The virtual community abides, and as in *The Truman Show*, surveillance and exhibitionism are drained of their negative connotation and made acceptable in the new media landscape. Likewise, with music and home recording or editing programs we "re-mix" or "mash up" sounds to create new sensations. And a kind of "self-publishing" or "re-imagining" co-exists with the ability to electronically share with and distribute to others. The following list outlines the key elements "Aggregate Spectating:"

Portability. Fifth and sixth generation audiences experience entertainment and information content "on demand." They are more comfortable with the concept of "nomadicity" than any previous set of consumers. A member of these generations may elect to never be "out of touch," to remain in a constant state of alert to new sounds and images, indeed, may feel disenfranchised if he or she is unable to maintain what was once called "remote access." Now it is just "access." The fifth and sixth generation aggregate spectator is really a structure unto himself or herself. They send and receive transmissions wherever they wish. The personal mobile device has made him a creator/sender/receiver/projector of media. And with a GPS, you're never lost. Perhaps a downside is, though, you can always be found.

Interruptivity. A new sensitivity to management of one's time and its apparent focus or concentration on new media forms is demanded by both audience and producer. The ability to be connected or disconnected at will is a hallmark of most portable devices. DVDs chop films into "chapters" with subtitled headings that allow spectators to skip to segments for re-viewing, unimpaired by the need to watch what the filmmaker created as orientation to those scenes in prior sequences. Music is segmented into "singles" for us now, by choice or design, or pre-determined listening segments, to allow a choice of purchase.

Driven by Multiple Activity. These generations are comfortable with the concept of "multitasking." Multitasking demands both an intellectual as well as physical or physiological comportment. The portabilization and segmentation of media space has increased the opportunities for multitasking to envelop a variety of media and communication tasks. Likewise, the ability to follow and engage multiple story threads is an attribute to be expected of fifth and sixth generation spectators. This is true whether you engage in video games or multithread TV series like *Luck* or *Game of Thrones*.

Customization and Selectivity. With the multitude of program and platform choices now available, it becomes necessary for some filtering of the consumer options amidst the media torrent. Fifth generation audiences became comfortable with "artificial intelligence" inclusions in software or affixed to user-profiles that both disallow certain messages to come through and give preference to others based on past uses/purchases or stated interests. The culture of RIYL (Recommended If You Like) persists on several platforms where we have created a history of preferences regarding media and consumer choices. Fifth and sixth generation spectators need to manage the apparent freedom of choice unleashed by new technologies, along with the tension created by those same technologies that can also foist a control or containment of those choices.

In closing, let me propose some stances and interrogatories regarding "How to Perceive/Appreciate Yourself as the Aggregate Media Listener-Spectator:"

• Embrace the role of purist, but not elitist.

• Remember that there is no such thing as an "old" movie or "old" music—only films you haven't seen before, or songs you haven't heard before.

• Understand the detachment from mechanical means of reproduction to digital ones; how is the "authentic" newly understood? What does it mean to "scratch the surface" of a record, or "distress the film"?

• Are there unique the challenges today in managing the quality of the spectator/listener experience? What are the social and cultural implications of managing the spectator/listener experience?

• Consider the ramifications of "free" or "paid" media and the ethical imperatives that accompany an abundance of choice today as a listener-spectator.

We're watching and listening to a wonderful world out there.

PART II:

DREAMS OF THE FUTURE: MODERN INTERPRETATIONS OF CLASSICAL LEGACIES

CHAPTER THREE

DANTE, DAMNATION, AND THE UNDEAD: HOW THE CONCEPTION OF HELL HAS CHANGED IN LITERATURE FROM THE INFERNO TO THE ZOMBIE APOCALYPSE

ISABELLE WHITMAN

Hell fascinates us. In literature and art, it provides ample opportunity for description, imagination and social commentary, and can even serve as revenge fantasy. Over time, the portrayal of hell in literature has evolved with society. As we reevaluate our ideas of sin, redemption and the afterlife, literature reflects the changing fears and anxieties attached to such concepts.

Dante is the godfather of the hell narrative, and almost every version of hell in Western literature owes something to his *Divine Comedy.* Dante's vision is based on Christian dogma, and his hell itself is meticulously detailed, organized and sin-specific, meaning that the punishments fit the particular sins. Although Dante's hell is in the afterlife, the tortures are predominantly physical ones. Dante uses imagery and characters from the Bible as well as his own imagination, and his hell is paradoxically not only in the hereafter, but physically located in the earth as well. The poet Virgil takes Dante on a guided tour through the underworld, which is literally under our world, and when they reemerge later, it is to climb a mountain leading to Purgatory, also located simultaneously on earth and in an afterlife. This lends the epic a sense of tangibility, that hell is closer, and more attainable, than one would like to imagine.

Dante adds another touch of realism when he incorporates actual people from thirteenth century Florence. He surely derived a fair amount of glee when, exiled and devastated, he managed to send his enemies to hell for all eternity, if not literally, then at least literarily, and judging by the *Comedy*'s considerable popularity, for almost as long. His epic work serves as both warning and revenge.

Often striking to modern readers of Dante, the order of severity of the sins and their consequent punishments seems reversed; the more animalistic sins, such as lust and gluttony, are punished less severely than the more human sins, such as avarice and treachery. At the time of Dante's writing, it was generally accepted that God was more forgiving of those who were simply too weak to resist their impulses than of those whose trespasses were calculated. He was much less merciful on those who used their capacity for reason to commit acts of sin. Therefore, the sinners who gave in to lust, even when it resulted in betrayal, such as Paolo and Francesca, are not punished for treachery but for licentiousness. Judas, however, is punished for the actual treachery, as his betrayal of Jesus was not an uncontrollable animal impulse but the result of a careful evaluation and subsequent choice. Divine justice in the *Divine Comedy* punishes acts of violence against others more severely than it punishes a mere lack of self-control.

Because they were unable to stop themselves from committing their sins in life, the sinners of the *Inferno* are damned to repeat them for all of eternity. They remain incapable of stopping themselves, as though addicted to their sin. Whatever vices they indulged in happily in life, they are forced to repeat endlessly in death: the gluttonous continue to indulge in gluttony, gobbling up filth and excrement the way they devoured food on earth; the violent continue to fight and tear one another limb from limb, unable to control their wrath; even the lustful, paired for eternity with the object of that lust, are unable to separate from one another, blown around and around, together, in perpetuity. This repetition itself is of course psychological torture, as the damned know all too well that they have no hope of ever escaping or even being momentarily relieved of their suffering; they know, too, that they are the cause of their own suffering and so have damned themselves to their own hell. The idea that hell is a place of one's own making is one that will be repeated throughout Western literature.

Jean-Paul Sartre, a French existentialist, did not find a medieval underworld of torture and demons compelling. He wrote his one-act play, *Huis clos,* in 1944, at the close of World War II, after being captured and held prisoner by the Germans; he knew firsthand the horrors humans can inflict upon one another. The overly dramatic wringing of hands and gnashing of teeth usually conjured up in descriptions of hell were a distraction to the real human suffering he saw and questioned on earth. Furthermore, Sartre seems to find the former preferable. His version of hell, from *Huis clos*, is often condensed to its most famous quote: "*l'enfer, c'est les Autres.*" (Sartre 93). Here, like in the *Inferno*, the damned have damned themselves, and part of their punishment is the knowledge that they are themselves responsible for their fate. In this grim imagining, however, there are no pitchfork-wielding demons, bodies in flame, rivers of excrement, or frozen lakes of despair. Instead, Sartre's characters torment one another psychologically, in a hotel-like room. Each character must admit his or her sin to the others, and as the play unfolds each realizes that the others are there specifically because they have the ability to afflict one another in a way appropriate to their earthly sins. As they gradually discover, each of them is the torturer-executioner, the "*bourreau.*" The play has only four characters, the three newly dead -- Garcin, Inès, and Estelle -- and the bellhop who brings them to the room. The three have never met before, and it soon becomes apparent that the reason they are placed together is that each has what another one wants, and each is unwilling or unable to yield that. Garcin, damned for both his cowardice in war and cruelty to his wife, desperately needs Inès to believe that he was not a coward but a conscientious objector, yet she will not. For her part, Inès, here for her role in a sadistic adulterous lesbian affair with her cousin's wife, is obsessed with Estelle, who has no sexual interest in women. Estelle is an infanticidal narcissist who drove her last lover to suicide when she drowned their child, and she requires constant male validation, which Garcin is unable or unwilling to provide her. It is never expressly stated in the play that the characters are in hell; it is only their assumptions that they are, and it has been conjectured that they are perhaps in Purgatory instead. Whatever the case, they all believe this place to be hell, and do not really question their place in it. Garcin, Inès, and Estelle know they have sinned, and accept, if unhappily, that hell is where they belong. It is possible that they may have the power to absolve, even free, one another, but again, this is not expressly stated and the audience may draw its own conclusions. Regardless of whether this power is assumed correctly or not, the point is moot, as none of them have the capacity to perform such a merciful act. Like Dante's sinners, they find they are unable to stop themselves from inflicting mental cruelty upon one another, repeating *ad infinitum* their earthly acts.

When Garcin finally comprehends, fully, the power Inès has over him and realizes she is the only one who can absolve him and that she never will, he begs for the tortures, the sulfur, the flames, and the physical agonies of a Dantean inferno. The irony, of course, is that when the door opens, Garcin finds himself unable to leave. Inès still believes he is a coward, and it is this very belief which tortures him. Although he knows she is correct, and has seen him as he is, he begs her to tell him she does not think he is cowardly. She is either unwilling or unable to see him as anything but a coward, and she even derives a sadistic pleasure from the deprivation, despite knowing he will avenge himself eternally upon her. Inès, like the two damned with her, has created her own hell, and like the sinners from the *Inferno*, is still addicted to the same sins she found herself unable to escape in life.

The title *Huis clos*, most often translated in English as *No Exit*, is a reference not only to the room, or to hell, but to the inescapability of the self. Although much is made about the lack of mirrors, the characters will never escape from their own reflections: the reflections they find in one another's eyes. They have each discovered the others' true selves, and once exposed, will never allow each other to hide again. Furthermore, the lack of mirrors symbolizes the loss of self-image; for eternity, they can see themselves only through the eyes of the others. Garcin, for instance, can no longer romanticize who he is and what he did, but is forever doomed to see himself as Inès does, cowardly and weak.

In the very first scene, before the women arrive, Garcin begins to understand the significance of this relentless new perspective when he notices that the bellhop never blinks, never closes his eyes. There are no windows and the lights do not ever go off. "*Les yeux ouverts. Pour toujours. Il ferait grand jour dans mes yeux. Et dans ma tête.*" (20) Garcin will be forced to look at himself, from the inside out, forever.

Sartre was not the only modern writer to take the fire and brimstones out of hell. C.S. Lewis, in his short work *The Great Divorce*, conceives of hell as a separation from God, not a place of eternal torture. Although those in this hell are suffering, their pain comes neither from physical torments nor from psychological ones. Those who find themselves in this place are lost, but no matter what sins they committed in life, the greatest is not knowing God, and so their punishment is to remain ignorant of Him eternally.

Lewis draws from *The Divine Comedy*; the book is presented as a dream the writer has, wherein he is taken on a tour of hell and heaven. Hell is a grey, bleak town in perpetual twilight, and its residents, called Ghosts, are cranky, cantankerous and querulous fools who know not the joys they miss. They have no physical needs, and can move about freely through the town, mostly farther and farther out and away from one another. They become isolated not only from God, but from one another, and any companionship they could offer. The narrator is invited to go with several Ghosts on a tour bus to Heaven, an Eden-like paradise full of light, radiant beings known as Spirits, where the Ghosts are given the option to stay. Only one does so, and only after being forced. When he is "saved," however, he transforms from a Ghost into a Spirit and gallops off to the sound of an angel chorus, never looking back at his fellow hell-dwelling Ghosts but apparently so filled with joy he has no more thoughts for his companions. Later, a saved Spirit woman meets her unsaved Ghost husband in the woods. He has become a small, withered, dwarf of a man, with a shriveled, puppet-like creature on his arm called the Tragedian. He longs for her to be aghast, to weep for him, but she has no pity for his misery, since he has brought it upon himself. She tells him she is "*in* love" now and that she was not truly in love with him before, as one "cannot love a fellow-creature fully" until one loves God (Lewis 84). This must be true, for if she had loved him, she surely could not bear to see him suffer so. As it is, she remains utterly unmoved as he shrinks hideously and miserably into nothing. He disappears altogether, literally reduced to nothing by his wretchedness and her indifference, and she wanders off singing. The narrator is shocked by this and asks his guide if it is not true that "the final loss of one soul gives the lie to all the joy of those who are saved" (110). The guide answers that if the Spirits felt sorrow for the suffering of the Ghosts, then their own happiness would be compromised, which would be akin to allowing the damned to "blackmail the universe" with their refusal to be happy on any but their own terms, i.e. without being saved or knowing God (111).

One who would agree with the narrator of the *Divorce* is Alex Hergensheimer, the hero of Robert Heinlein's 1977 novel, *Job: A Comedy of Justice*. In an apocalyptic world, Alex, a minister in a fundamentalist Christian church in an alternate universe United States, falls in love with Margrethe, a Pagan and worshipper of Odin. For unexplained reasons, dimensions constantly shift, placing Alex and Margrethe in strange new worlds unpredictably, until the eventual Rapture. Alex sees the shifts as portents of Revelation and tries desperately to bring Margrethe to a state of grace, while she believes it is Loki bringing about Ragnarok, the Viking version of end times. They are separated, each to his own apocalypse. Alex, who is saved, goes to heaven, but finds it no paradise without Margrethe, telling St. Peter that "an eternity here without her is not an eternity of bliss; it is an eternity of boredom and loneliness and grief." (312) He begs to be thrown out of heaven to regain his love, and St. Peter acquiesces, literally tossing him out and over the edge.

Alex falls into hell, which is a bureaucratic mess much like heaven, and finds that Margrethe is not there either. She fought bravely in Ragnarok, and went to Valhalla. Alex finds Loki and Odin and eventually persuades them to ask Margrethe if she would like to leave Valhalla and rejoin him, to which she immediately agrees, as she too finds her paradise lacking when her beloved is absent.

This idea is in sharp contrast to Lewis' Spirits, blissfully untouched by the suffering of the damned Ghosts, and especially from Dante's heaven, where the saved may look down upon the suffering in hell and rejoice at their well-deserved agony. Heinlein's heaven and hell are both more human than the others' version. Not only are they organized like government offices, but the souls therein behave the way humans do on earth: they are bored, annoyed, curious, lazy, or lonely. Most significantly, they are still sympathetic, empathetic even, to those whom they loved in life.

All of these hells, whether directly Christian or vaguely agnostic, have at their center a structure, the workings of an unseen hand, a hierarchy or system. But what if one were to remove that presence? Film director George Romero does just that. If Dante perfected the hell fiction, Romero invented the zombie genre. Combining the gore and carnage of Dante's lower circles with the "hell is other people" concept of Sartre, the zombie apocalypse is the contemporary literary incarnation of hell. Not only is hell other people, but it is created by other people as well: zombies are not naturally occurring, nor are they sent from God as a punishment for sins. They come from military experiments gone awry, nuclear disasters, or biochemical warfare. Whatever the details of the cause, it is always humans who create it.

There is no God in the zombie apocalypse. Romero has removed the guise of religion, and has also abolished the demons or bellhops. There are no guides, no neat circles, and no enclosed rooms. People do not die sinners and get sent to hell; they live, and hell is sent to earth. One of the most chilling lines, and the tag line for the movie itself, comes from *Dawn of the Dead*: "When there's no more room in hell, the dead will walk the earth."

Zombie narratives are hugely popular in the film and literature of pop culture right now. "More than any other monster, zombies are fully and literally apocalyptic ...: they signal the end of the world as we have known it" (Paffenroth 13). Zombies reflect modern fears of science and technology, of our fellow man, and nuclear or chemical terrorism.

In the zombie apocalypse genre, hell is no longer either an afterlife or a realm of supernatural demons, but is instead right here on earth, and the torments, while often grisly, are mostly psychological. There is of course the horror of being eaten alive by a humanoid enemy, but the more harrowing moments are when a character must witness a loved one "turning" and inevitably becoming inhuman, an ordeal made more horrifying by the realization that one must kill, or be killed by, those who were once family or friend.

Compounding this horror are the other survivors. Unlike in *Huis clos*, the ones who are still around are not there for any discernible reason: some are lucky and have banded together; some are smart or prepared; some are just vicious enough to live through almost anything and use the anarchic situations as a means of indulging cruel or sadistic whims. The following scenario is so common in zombie films and literature as to be cliché: a survivor or group of survivors stumbles across others who appear friendly, but then turn out to be as much of, if not more than, a threat as the zombies themselves. In *Night of the Living Dead*, it is the wandering group of strangers who kill the would-be hero. In *The Walking Dead* series it is the Governor who leads the makeshift town and attacks the prison. In *Day of the Dead* it is the Army men threatening rape and murder of the scientists. The theme is always the same: *we* are the enemy, the monster; *we* are corrupt and cannibalistic; *we* are rotten, whether the putrefaction is literal or metaphorical.

Phalanx, a bogus drug marketed as a vaccine against "African rabies," as the zombie virus is called in Max Brooks' *World War Z*, is a cynical indictment of capitalism and the pharmaceutical industry. The novel is written as a compilation of interviews done by the United Nations a decade after the start of the zombie apocalypse. Breckinridge Scott is an opportunist, a marketing genius who sees in the zombie outbreak a chance to make literally millions of dollars by selling false hope. He explains advertising is not the selling of products, but the commodification of fear. "People selling their products? No. People selling the fear of you having to live without their products ... Fear of aging, fear of loneliness, fear of poverty, fear of failure ... Fear sells" (Brooks 55). The sham vaccine helped the virus to spread, as people thought they were safe. Once again, man's greed is his downfall.

In zombie films and literature, society as a whole is punished for its collective sins, whereas in previous hell literature each character is punished for his or her individual sins. These societal sins usually include racism, sexism, classism, greed, consumerism, warfare, and other violence. In *Night of the Living Dead*, Romero's first film, the hero is a black man. He fights bravely, keeps his head in terrifying situations, thinks quickly and tries to save all he can, and is shot in the head by a careless crowd of white men and their dogs. They did not know who he was, they had no reason to think him a zombie, and they remain ignorant and callous to the end, using meat hooks to put his body on a pyre of corpses, human and not. He would probably have been their salvation from the violent end we assume they will meet. Casting a black man as the heroic lead in a story of violence, chaos, and human self-destruction was no small statement in the years immediately following the assassinations of prominent black leaders Malcolm X and Martin Luther King, Jr. The whites who kill Ben are reminiscent of a lynch mob, with their shotguns and dogs wandering through woods and fields. When they destroy him, they are destroying their own hopes of survival, yet they are no less ignorant of this than were the lynch mobs of the 1960s.

In the second of Romero's pentalogy, *Dawn of the Dead*, the most sympathetic and resilient character is a woman, Francine. Women in horror films are often depicted as skittish, useless and dumb. Not so in Romero's and subsequent zombie works. Francine is smart, resourceful, and in fact stronger than her male counterparts. When they are thrilled to discover a shopping mall (still a novelty in 1978), she is disgusted by their materialism. Only Fran sees the uselessness of jewelry, clothes and other possessions. This is Romero's way of thumbing his nose at the dismissive stereotypes of vapid female characters found all too often in popular culture. Here, it is the men who shop until they literally drop, and the woman who is exasperated by their blind greed.

Another cliché Romero dismantles is the "buddy movie" trope: two men, one black and one white, are friends, but invariably the white man is the lead and the black man is the humorous, and often expendable, sidekick. Again, Romero surprises. There is a strong male friendship between Roger, a black man, and Peter, a white man, but it is Roger who is the lead, and Peter the expendable sidekick: in this case, Peter is bitten by a zombie and will slowly die, only to come back as one of them. In a harrowing scene, Roger alone stays with his friend, the two men knowing what will happen and what must be done, and finally, Roger having to have the strength to shoot and kill Peter when he dies, and reanimates as a monster. Then, Roger must battle through the hordes of undead to rejoin Francine in a helicopter and escape.

It is significant that Francine is the one piloting the helicopter for the final escape. Earlier in the film, when Francine reveals she is pregnant, the men immediately discuss ways of aborting the pregnancy, without acknowledging her. They had already assumed control of her and her body. She vehemently protests this objectification, and Roger, who only now realizes she is an equal, decides she should be taught to pilot the helicopter and that she should carry a gun at all times. In the end, this saves him, and probably humanity, as the two of them escape in the helicopter she can now pilot, presumably to make a life for the unborn baby.

The ambiguous ending is the most optimistic one in all of Romero's works, but it is not the one he intended. In the original, no one survives: Roger shoots himself rather than be eaten, and Francine decapitates herself with the helicopter's blades when trying to escape. The original final shot was the blades slowly running down, indicating they would not have had enough fuel to escape anyway (Paffenroth 63).

Apart from the individual horrors of hell, one of the defining characteristics of any hell, whether Biblical, allegorical, or literal, is the idea of eternity. Nowhere is this better illustrated than in Robert Kirkman's ongoing graphic novel series, *The Walking Dead*. As Kirkman says in his 2003 afterword, the most frustrating part of zombie movies is that they end. The audience knows that there is more, but of course a film cannot last forever. One is always left wondering, as Kirkman says, instead of allowed to "follow the end of the world to its natural conclusion." And so his series is subtitled: *A Continuing Story of Survival Horror*. It has proven to be wildly successful, publishing monthly since 2003 in comic form, and a popular television series now in its third season. But why are they so popular? Why do today's audiences identify so well with the zombie apocalypse?

Although the popularity of the zombie as a present day bogeyman is fairly recent, Marina Warner suggests that the concept of a zombie was present as far back as the Middle Ages. One of the earliest mentions of zombie-like beings in Western literature comes from none other than the godfather of hell himself, Dante. In Canto 33, Fra Alberigo, the murderous monk, claims Branca d'Oria is there in Cocytus and has been for years, but Dante argues that this is not possible, as he has just seen him alive on earth, eating, drinking and sleeping. Alberigo explains that some sinners are so evil the devil may claim their souls before they even die, leaving their emptied bodies to roam the earth (Dante 411). The great poet had "imagined the possibility, but did not have a word for the soul-less, hollow man left in the world above" (Warner 125). Since then, we have not only the name, but an entire genre.

The fear of soullessness seems a natural side-effect of a soulless culture. As we become more materialistic, we become less compassionate. We distance ourselves from one another by interacting through machines instead of face to face. We become jaded and desensitized as we are fed a constant stream of violence in news and other media sources.

As each generation scoffs at the fears and inhibitions of the last, those fears change and shift focus. Maybe examining them from the safety of literature helps us to overcome them. Zombie fictions force us to confront those fears in a visceral and relatable setting. When we spend several hours getting to know characters, we identify with them, and it is shocking to see them killed. We can imagine ourselves in their place. We see our worst fears played out in Technicolor, and yet we survive them, perhaps conquering them in the process. Whatever the case, our fascination with death, afterlife, punishments and damnation seems to be in no danger of diminishing.

References

Alighieri, Dante. *The Divine Comedy*. Trans. John D. Sinclair. New York: Oxford UP, 1961. Print.

Bringsjord, Selmer. "The Zombie Attack on the Computational Conception of Mind." *Philosophy and Phenomenological Research* 59.1 (1999): 41-69. Print.

Brooks, Max. *World War Z: An Oral History of the Zombie War*. New York: Crown, 2006. Print.

Gunzberg, Lynn M. "Down Among the Dead Men: Levi and Dante in Hell." *MLS* 16.1 (1986): 10-28. Print.

Harrison, Robert P. "Comedy and Modernity: Dante's Hell." *MLN* 102.5 (1987): 1043-061. Print.

Heinlein, Robert A. *Job: A Comedy of Justice*. Toronto: Del Rey, 1985. Print.

Kirkman, Robert, writer. *The Walking Dead*. Art by Charlie Adlard and Tony Moore. 2 vols. Berkeley: Image Comics, 2003. Print.

Lewis, C.S. *The Great Divorce*. San Francisco: Harper, 2001. Print.

Milton, John. *Paradise Lost*. London: Penguin, 2003. Print.

Paffenroth, Kim. *Gospel of the Living Dead George Romero's Visions of Hell on Earth*. New York: Baylor UP, 2006. Print.

—. *The Heart Set Free: Sin and Redemption in the Gospels, Augustine, Dante, and Flannery O'Connor*. New York: Continuum International Group, 2005. Print.

Pike, David L. *Passage Through Hell: Modernist Descents, Medieval Underworlds*. Ithaca, N.Y: Cornell UP, 1997. Print.

Romero, George, dir. *Dawn of the Dead*. United Film Distribution Company, 1978. DVD.

—. *Day of the Dead*. United, 1985. DVD.

—. *Night of the Living Dead*. Walter Reade Group, 1968. DVD.

Sartre, Jean-Paul. *Huis clos, suivi de Les mouches*. Paris: Gallimard, 1947. Print.

Turner, Alice K. *The History of Hell*. San Diego: Harcourt, 1995. Print.

Warner, Marina. *Fantastic Metamorphoses, Other Worlds: Ways of Telling the Self.* New York: Oxford UP, 2003. Print.
Whitaker, Thomas R. "Playing Hell." *The Yearbook of English Studies 9* (1979): 167-87. Print.

CHAPTER FOUR

NOMINA NUDA TENEMUS: JONATHAN SAFRAN FOER, FINDING MEANING WITHIN EMPTY NAMES OR (RE) CONSTRUCTION OF DECONSTRUCTION

ASHLIE M. CONTOS

“□.”

“ICELAND, 22:13:36. April 11, 2006, Variable Point.”
“‘{I want to love and be loved.}’
‘□’
‘□’
‘↓’
‘↓’
‘□’’’.[1]

What do these seemingly random symbols mean? At first glance, nothing, but within the narrative worlds created by Jonathan Safran Foer, each of the above symbols carries meaning of great depth. Foer’s characters are set

---

[1] These symbols appear in the following works, respectively: “A Primer for the Punctuation of Heart Disease” 1, *Extremely Loud and Incredibly Close* 115, “About the Types Face Not” n. pag., and “A Primer…” 9.

upon by trauma of various degrees: death, alienation, aphasia, loss of a parent/spouse/son in the September 11 terrorist attacks, heart disease, the Holocaust, "the prison-house of language,"[2] loss of a pregnant girlfriend in the bombing at Dresden, loss of a pregnant sister at Dresden and the list goes on. In writing about events that remain delicate, even over seventy years later (i.e. the Holocaust), Foer addresses the horrors of these occurrences and does so by creating new expressions for these indescribable experiences. Foer continually manipulates words, emoticons, visual media, and narrative structures to demonstrate language's limitations, especially when it fails as a means of conveying suffering, grief, angst and death. Despite language's inadequacy, Foer strives to address traumas of the past through an inventive, twenty-first century mode–one that relates to the current aesthetics of his readers. Trauma demands a new aesthetic exhibition in order to address and connect with a twenty-first century readership. Rachel Greenwald Smith observes that, "[w]e look to trauma to offer opportunities for new ways of seeing and new ways of thinking, and we expect this upheaval in sensory, emotional, and intellectual experience to transform literature on the level of form" (154). Foer transmutes his literary works by emphasizing the inexpressible nature of violent events; while his characters may struggle with the traditional narrative of I-tell-you-the-reader, they endeavor to more fully express their emotional interior by showing-you-the-reader through images or lack thereof [i.e. blank pages which represent the ineffable]. Though the narrative incorporates both linguistic and pictographic communication, it is language and words that ultimately bind characters together and alleviate their emotional burdens. While language falters at times, it ultimately remains as mankind's most effective mode of communication in peace and in turmoil; language serves as the bond that binds us together. Thus, unlike many deconstructionist conclusions–that linguistics as a form of communication is unstable and meaning is therefore unattainable–Foer acts as a post-postmodern model of (re)engagement as he explores resolution and existential meaning even in the face of tragedies of human existence.

Foer's (re)construction of post-modern theories addresses the void left in deconstruction's wake and creates a new kind of bridge between signs and meaning, and the creation of both individually and jointly. I propose that J. S. Foer is a post-postmodernist, meaning that through the fluctuation of sentiment and irony the author connects to the reader through the text

[2] As defined by Frederic Jameson in The Prison-House of Language: A Critical Account of Structuralism and Russian Formalism.

thereby generating engagement, feeling and meaning in spite of deconstruction's dismantling of words and post-modernism's insensibility. Nadine Fessler notes in her article, "To engage in literature," that while contemporary literature recurs, "outdated concepts such as authenticity, ethics, engagement and aesthetics...[it] does not dismiss postmodern ideas, but rather incorporates those ideas, though utilizing them for completely different outcomes." Foer blends modernist ideals of universal meaning and sincerity with post-modern irony and persiflage in order to create a metamodern novel reminiscent of twenty-first century socio-cultural dynamics. This chapter will briefly define post-postmodernism, explicate Foer's concurrent use of semiotic-play and earnest dialogue in "A Primer for the Punctuation of Heart Disease" and *Extremely Loud and Incredibly Close*, and illustrate how he produces meaning and connection through these texts, thus moving him into a new breed of writers who are exemplifying the post-postmodern paradigm.

The term post-postmodernism or metamodernism first appeared in the early 2000s, and it is presently gaining more recognition and significance due to websites such as Notes of Metamodernism and publications from Rodopi's Postmodern Studies series. Just as post-modernism was a reprisal of modernist theories, post-postmodernism reacts against or due to post-modernism's influence over cultural and literary studies in the 1960s and 1970s. Though deconstructionist theories are considered outdated, it seems there has been a void in literary critiques since post-modernism's decline. Scholars are now researching the new socio-cultural shift into what they are defining as post-postmodernism or "po-pomo" (Hartness 2). According to Katelynn Schoop, po-pomo continues the deconstructionist theory that definite meaning cannot be achieved through language yet simultaneously proposes that "fiction now searches for the experience of emotion" (1). Schoop's conception of this new movement is in accordance to the definition promoted by the Notes on Metamodernism's editors, who define metamodernism as a philosophy that, "oscillates, swings back and forth, between the global and the local; between concept and material; between postmodern irony and a renewed modern enthusiasm...it strives for sincerity without lacking humour, it engages precisely by embracing doubt" ("Discussing Metamodernism"). It is because of and through doubt that we are forced to reach out of ourselves, our own subjectives, to discover and embrace the life we have before us and address our fears and anxieties. Thus, post-postmodern authors shift their literary discourse toward broadening the reader's subjective so that it can connect with another's subjectivity, thence "[e]mpathy becomes the site at which fiction's social purpose can be realized" (Schoop 2). Nicoline Timmer, the most extensive author on po-pomo, agrees with Schoop and expounds on the concept of the subjectivity and connection through empathy, stating that within this new generation of authors there is a movement toward "expression of feelings and sentiments, a drive towards inter-subjective connection and communication, and also a sense of 'presence' and 'sameness'" (13). Presence and sameness within po-pomo contrasts deconstruction's presence and absence; whereas deconstruction finds a void between presence and absence—binary opposites have no identifiable means of reconciliation—po-pomo looks to commonalities in order to produce meaning. Robert L. McLaughlin narrates the change from post-modernism to po-pomo as "being inspired by a desire to reconnect language to the social sphere...to reenergize literature's social mission, its ability to intervene in the social world" (55). The consensus seems to determine that these new young authors begin with David Foster Wallace and Dave Eggers in the late 1980s and early 1990s, and I would argue that they are now followed by millennialists such as Jonathan Safran Foer. Because literature typically speaks to or depicts the human condition, many studying literary theory today do not consider post-modern theories adequate in examining the subjectivity of mankind.[3] Some literary scholars are returning to theorists such as Baktin and Peirce in order to regain a sense of meaning and connection from and within literature. Katrin Amian studied Foer's *Everything is Illuminated* through Peirce's concept of 'consensus'; whether Foer has studied Peirce's works is uncertain, but Foer too builds his literary works off of what he calls "collective creation" (Amian 189).[4] Amian describes this term as Foer's description of "the larger cultural, political, and societal processes of (self-)reflection and creative invention to which he sees his novels contribute and respond" (189). A focus on the "larger cultural, political, and societal processes" is what makes literature pertinent and influential to its reader. Therefore, meaning is not solely dependent on words and signifiers, but ultimately, through our ability to feel, recognize that emotional state, and express our sentiments therein, finding the connection to one another and to meaning. These connections are built across the inner-subjective of individuals to attain an inter-subjective based on

[3] Nicoline Timmer quotes many scholars who question post-modern critiques, considering them inadequate, including: Lois Tyson, George Butte, and Uri Marolin.

[4] Amian quotes Foer from a reading he did while at Buchhandlung Bittner on Wednesday, 21 September 2005, in the Königin-Luise-Schule, Albertusstr. This reading was not recorded, so the only source I have to this term "collective creation" is from Amian's relation of it.

common emotions and experiences. Language functions as a means for individuals to serve and contribute to those around them, and despite its own instabilities, it remains as mankind's greatest endowment.

As Foer and other post-postmodern authors strive to shape this new movement, it is noted by scholars and critics as a duality between modernist and post-modernist tendencies. Foer continues this duality between the two eras, yet he retains many aspects of deconstructive and semantic-parody, which is best seen in "A Primer for the Punctuation of Heart Disease." If signs "denote a basic element of communication," which we typically relate to words, Foer extends this definition through a discourse of textual symbols and visual media. "A Primer..." is littered with text-signs that the protagonist, Jonathan, employs to express his Jewish family's repression. If language is an inept form of communication, then it fails even more in light of tragedies that remain unutterable; how can one convey the horrors of the Holocaust or the fear of loving because of the loss created during the Holocaust? Foer here deviates from traditional linguistics as Jonathan narrates his Jewish family's aphasia through silence marks (□, ■) and expressive symbols (□, □and {}). As Jonathan tries to explain his family's silences, he simultaneously reveals the repressed trauma, as when his grandmother drives him to a piano lesson and says to him, "'I hope that you never love anyone as much as I love you¡'"[5] (Foer 2). According to Jonathan, this almost unnoticeable punctuation mark "¡" indicates a whisper. Jonathan recalls that they were listening to the audio version of the Shoah when Grandmother tells him this odd obiter dictum, but he does not yet mention that his family is Jewish or that his grandmother survived the Holocaust. It is not until eight symbols later that Jonathan alludes to his family's loss because of the Shoah. Jonathan introduces the reversible colon (::), which connects "the two halves of the sentence [to] explain each other" (7). He begins by describing his familial relationships, but he then includes more personal sentiments:

> "1938 to 1945::□.
> Sex::yes.
> My grandmother's sadness::my mother's sadness::my sadness::the sadness that will come after me" (7).

---

[5] It is interesting to note that this same line is used in Extremely Loud and Incredibly Close within a conversation between Grandmother and Oskar, her grandson. Again, a grandparent who suffered trauma from a previous war tries to guard the next generation from similar painful emotional rifts.

The estrangement builds as Jonathan describes the last symbol, {}, what he calls the "should-have brackets" (8). This sign is placed around the sentences that he or his family do not actually say but should have, and he demonstrates this in a conversation with his father that ends with:

> "'◎□❖◎○◎◦□◆⊙●'
> '■'
> '{I love you.}
> '{I love you, too. So much.}'" (9).

Jonathan's desperation to construct a sense of dialogue from the silences within these conversations mimics Foer's attempt to fill the gaps between author, reader, and meaning due to post-modern disillusionment. The semiotic experimentation does not end in nothingness, as deconstructionists would conclude; it ends with Jonathan seeking to create out of the nothings within his familial conversations. Foer ends the short story with the post-postmodern turn toward a desire for a connection between people, between the subjective of the characters and that of the reader. Jonathan ends by trying to identify with his family's version of the should-have conversations just as the reader endeavors to interpret and understand Jonathan's emoticon language. Because the "should-haves" represented by Jonathan are from his own subjective, the reader cannot know how similar or dissimilar these conversations are interpreted by his father, mother, or brother. As Jonathan invents these emoticon dialogues as a textual representation of his familial angst, the reader must imagine what Jonathan himself intends these symbols to mean as the short-story gains momentum and Jonathan refrains from explaining every symbol and its use. Jonathan's emoticon incongruity reflects the post-modern theory of the slipperiness of language, but his resolution demonstrates the metamodern shift toward reconciliation between expression and meaning due to common experiences that bind humanity together. Jonathan tries to reach out of himself as he "imagine[s] their [his family's] should-have version. I sew them together into a new life, leaving out everything that actually happened and was said" (9). Although words may fail Jonathan and his family, Jonathan seeks to translate the silences within the familial repressions, just as Foer attempts to construct meaning despite deconstruction's assault on language and verbal exchange. Fessler observes that, "[w]hereas in postmodern literature, language's ability is subverted, meaning is deconstructed and truth is ever-displaced, Foer seems to seek strategies to overcome the limitations of language, while not ignoring them" ("Illuminating Language"). Thus, the semiotic banter seen between words and emoticons while tempered by Jonathan's yearning for emotional, earnest dialogue – between himself and his family and between himself and his reader – creates a literary rostrum of metamodern characteristics and ideologies.

Whereas Jonathan in "A Primer..." deals specifically with the turmoil among and within his family unit, Oskar in *Extremely Loud and Incredibly Close* eventually explores everyone's (himself, his family, NYC, the reader) vulnerabilities because the 9/11 terrorist attacks affected, at least, everyone in the United States. Oskar talks about his own trauma and feelings throughout the novel in the chapters he narrates, and though Oskar may not include everyone in his grief initially, by the end of the novel he comes to understand that he is not the only one suffering from loss. As Oskar journeys through his quest to find the answer to the mysterious key labeled "Black," he realizes that his subjective experience is not insular. Kristiaan Versluys astutely notes that all the men who assist Oskar in the "Black"-key mystery are also in need of help and connection, and she states that A. R. Black, William Black, Stephen Hawking, and the "renter" (i.e. Grandfather) are, "truly interlocutors, in that their sympathy for Oskar originates in the personal experience of suffering and deprivation." Their ability to empathize is due to their own experience with suffering. While words may prove to be clumsy in expressing the character's individual grief and sympathy, the point is that they traverse language's complications because the need to connect and feel is a greater concern than getting it exactly right. Therefore, as Oskar discovers that he can relate to others' experiences and they in turn can empathize with his loss, Foer petitions the reader(s) to find the bridges that connect us to the characters and in a larger scale to others. Natalie E. Couch notes that authors who employ a child-narrator, "ask their adult audiences to rethink the way they process information and therefore relearn in an untraditional manner" (12). Foer navigates the loss and fear created by the 9/11 terrorist attacks through the eyes of a child; an individual who is at a stage in life where one is already precocious, vulnerable and developing. If Oskar at times seems a "ludicrous creation" (Bissell 28) or "equal parts unbelievable and unbearable" (Siegel n. pag.) because of his emotional desperation and existential crisis, then Foer is daring to perform David Foster Wallace's description of a "[r]eally good work," one in which the author will "disclosure yourself, open yourself up in spiritual and emotional ways that risk making you banal or melodramatic or naive or unhip or sappy" (McLaffery 148-49). As Foer acts as a post-postmodern author, he shrewdly transitions from the post-modernist sematic play (photographs, Morse-code, inserted edit marks, empty pages of Grandfather's inexpressible thoughts and feelings, and phrases like, "I was already out of words when I met your mother") into post-postmodern literature with a child-narrator who is desperately seeking to understand the "Stuff That Happened to Me" (28, 52). This idea of trying to make sense out of one's experience is directly related to the post-postmodern concept of "Self as Narrative" as defined by Nicoline Timmer, who concludes po-pomo is about "how we try to make sense of who we are. The relevant question, therefore, is not: 'What is a human being?' (searching for fundamental essences), but: 'What does it mean to be me?'" (27). Foer tackles the question 'What does it mean to be me' under devastating circumstances, and he resolves the question partially by demonstrating our need of others and the necessity to share and connect with them. Foer reaffirms literature's influential power over the reader as he demonstrates the dynamics and effects within fictional relationships and how the fictional narrative represents and encourages engagement between text and reader – through emotional, visual, and intellectual intentness – and the reader and his/her own community.

Foer not only implies the need for meaning on an interpersonal level, he extends this necessity into an international stratum. At the beginning of this essay, metamodernism is noted to be a balance between global and local concerns; again, Foer demonstrates a mindfulness of problems both at home on American soil and those across the seas. Foer identifies the United States' victimization during 9/11, but he also references the bombing of Dresden and Hiroshima—strafes that created victims because of American involvement. Though Foer does not speak politically in the novel—Oskar is after all a nineyear-old child—he does implore the reader to become conscious of how we affect one another, interpersonally, inter-socially, and transnationally. Perhaps one of the most interesting details about the novel is that the protagonist's familyare German-Americans; Foer writes empathetically toward a nationality that sought to demolish his familial roots in World War II. Foer negotiates the interaction between offenders and victims inside and outside of his literature; an intercourse that is, "unthinkable in a postmodern world where we tend to experience the story through the victim's eyes" (Fessler n. pag). Foer chooses to identify with the German casualties of WWII by depicting the trauma and repression experienced by Oskar's grandparents, and consequently, how the family's suppression affects later generations. Versluys notes that by contrasting victims of different nationalities and different wars, *Extremely Loud and Incredibly Close* "universalizes grief" (n. pag.). While Oskar and the reader are repeatedly represented with the suffering every individual, family, community, and country experiences, Foer emphasizes the overarching event of grief that serves as the common denominator of all human beings. Post-postmodern theory seeks to overcome post-modernism and deconstruction's dismantling of language in order for authors and readers to once again reinvest in language and in our own ability to express our sentiments, thus leaving behind a certain cynicism and chagrin. Metamodern theorists recognize the need for a verbalization of the anxieties and restraints we feel due to the nature of language as well as the overwhelming sublimity of life's ordeals without the plunge into fatalism generated from deconstructive premises. Our investment in words and language must not be abandoned because of its very deficiencies. D. F. Wallace "explicitly calls for, and questions the possibility of, a kind of writing that could learn from the linguistic and aesthetic insights of the post-moderns while retaining the traditional meaning of fiction, which he describes as exploring 'what it feels like to be a fucking human being" (McAdams 5).[6] Foer takes this investigation of what it feels like to be a human being to the extreme as he analyzes human suffering; an emotion or occurrence that every individual will experience in life. He writes about a subject that is inherently universal, and he does so while bringing Oskar outside of his somewhat isolated subjective. Oskar has trouble understanding why anyone other than himself would have a reason to cry after the 9/11 terrorist attacks. When he visits the first location in the "Black" key mystery, Abby Black's home, she begins to cry during their conversation, and Oskar ponders on her reaction: "I thought, I'm the one who's supposed to be crying" (96). This self-absorbed melancholia reflects what Carlini describes as "an American narcissism unable to conceive of a trauma surpassing 9/11 in the wake of the attacks" (34). As Oskar chronicles his journey only a year after the death of his father in the 9/11 tragedy, it is not surprising that Oskar is self-consumed in his desolation. Yet he begins to realize how omnipresent grief is, and it is when he meets Mr. Black that Oskar grasps how truly connected we are to one another. While we all may not share the same suffering, humankind suffers together; thus while we may be separated by time, culture, ideologies, etc., we will always share a common experience of pain and suffering. Mr. Black explains war, music, money, and isolation to Oskar in the form of note-cards that he has cataloged his entire life; Mr. Black illuminates Oskar to the suffering that occurs to everyone in life. It is here that Oskar realizes that "[e]veryone could know what everyone else felt, and we could be more careful with each other" (163). Oskar comes to the

[6] This quote from David Foster Wallace can be found in the interview given by Larry McCaffery. "An Interview with David Foster Wallace." *Review of Contemporary Fiction* 13.2 (1993): 127-50.

heart of po-pomo criticism: empathy. Couch notes that the novel ends only "when individuals have been (re)connected with one another" (29). It is only when Oskar digs up his father's grave with the "renter" (i.e. Grandfather), buries the unsent letters and conciliates with his mother that Oskar can end his narrated chapters. Oskar sets aside his own grief so that he can reach out to his mother (who he has kept an emotional distance from until now):

> "I told her, 'It's ok if you fall in love again.'
> She said, 'I won't fall in love again.'
> I told her, 'I want you to.'
> She kissed me and said, 'I'll never fall in love again.'
> I told her, 'You don't have to make it up so I won't worry.'
> She said, 'I love you.'" (*EL&IC* 324-35)

Though Oskar struggles with expressing his anger, confusion and depression in the aftermath of his father's death, he finally bridges the linguistic divide once he is illuminated to the universal perplexity of expression, meaning and human emotion. Foer leaves postmodern disillusionment behind so that "language then, is a welcome problem, because once overcome it unites rather than diffuses" (Fessler n. pag.). Once Oskar overcomes his unvented feelings, he is then able to reach out and ease his mother's suffering due to their estrangement. Schoop notes that D. F. Wallace emphasizes the importance of the "externalization of one's emotional interior in escaping solipsism and forming real social relationships" (54). Within this passage, Foer follows Wallace's ideology of verbalizing one's feelings when Oskar finally opens up to her in a positive manner. Oskar empathizes with his mother and expresses his desire for her to be happy—or as happy as one can be after losing a spouse. Oskar develops from having "heavy boots"[7] to still having "heavy boots" but being able to engage and identify with those around him.

Oskar concludes the novel with the metamodern desire for an engagement with hope. For Oskar, it is a hope of a different past, one including instead of eradicating his father. Oskar's resolution in his final chapter, "Beautiful and True," reverses the order of his Stuff That Happened to Me memory book so that the 9/11 photograph of the falling man falls no more but seems as if he, "was floating up through the sky," which is followed by a list of subjunctive events that would not have occurred (325). "Dad would've left his message backward"; "He would've

[7] A term Oskar uses to describe his depression and sorrow from the loss of his father, Thomas Schell.

spit coffee into his mug, unbrushed his teeth, and put his hair back on his face"; "He would've gotten into bed with me"; and "We would have been safe" (325-6). Vermeulen and van den Akker note that while metamodernism shifts between modern and postmodern tendencies, it ultimately, "moves for the sake of moving, attempts in spite of its inevitable failure; it seeks forever for a truth that it never expects to find...thus [it] willfully adopts a kind of donkey-and-carrot double-bind" (n. pag.). Oskar rewrites "the worst day"; he creates a sense of fictitious happiness for the sake of producing a visual, though illusory, alternate reality for himself that he can witness through a physical object—a book (11). Oskar uses literature—his journal entries—as a means of documenting, working through and creating his personal expression of self. He testifies to his own trauma while reaching out to sense a possibility for a different reality. Smith sees that "art today might be the privileged seat of politics, given that 'politics truly means the creation of new possibility. A new possibility of life, a new possibility of the world'" (160). Oskar recognizes the possibilities before him to create a rewrite of history, of his personal history. By the end of the novel, Oskar is more than aware of his grief and the healing that has taken place because of his interaction with those around him, and he continues to act as his own therapist/author when he rewrites September 11. He hopes for the sake of hoping that his life could have been better, and the love he felt from his father could have gone on and on and on.

"[T]here is no center to our understanding of existence" (Tyson 249). This is where deconstruction leaves us. Yet founding post-postmodern authors look beyond the impotence of language and instead find meaning in emotion. David Foster Wallace writes in his short story "Brief Interviews with Hideous Men" that we choose our own subjective take on meaning, saying, "you can choose to be more if you want, you can choose to be a human being and have it mean something" (123). Foer seems to agree with both of these ideologies. Foer continues Wallace's existential notion in a literary endeavor so that literature can once again influence and contribute to its reader and the reader's society. It contributes to the human experience while acknowledging language's limitations. In his introduction to *The Diary of Petr Ginz*, Foer admits that "[w]ords are perverted by our histories (personally and globally), by context and tone of voice. Words are bad approximations" (vii). Yet Foer does not end with this nihilistic thought; he suggests that even after horrifying events like the Holocaust—which Petr Ginz did not survive—that language can have meaning, that it "must be reconstructed with an energy greater than that of its destruction" (x). Post-postmodernism seeks to reinvigorate language through the expression of sentiment whereby we might connect to one another through our sameness as human beings: love, sorrow, laughter, and suffering. If D. F. Wallace is correct, authors can choose to convey meaning and truth in their texts, and readers can choose to generate meaning from literature as they engage and respond to its message. Foer seeks to create meaning between his characters in both the works detailed in this chapter, and as I have argued, that meaning can be felt and known by the reader because of its universal nature. The reader can choose to connect and make meaning despite language's inadequacies, or as Foer has eloquently put it, "Readers are always running toward the same place when they open a book. We're trying to get to a greater understanding of ourselves…[w]riters are all running, too. They're running to meet their readers" (*Words Without Borders* 3). Foer makes the choice to run to his reader with literature of hope. Hope of bridging any dissonance or void within language in order to construct a common understanding with his reader and move into a post-postmodern paradigm of human connection and reciprocity.

References

Amian, Katrin. "Creativity and Consensus: Jonathan Safran Foer's Everything is Illuminated." Rethinking Postmodernism(s) Charles S. Peirce and the Pragmatist Negotiations of Tomas Pynchon, Toni Morrison, and Jonathan Safran Foer. Ed. Theo D'hean and Hans Bertens. *Postmodern Studies* 41. Amsterdam: Rodopi, 2008. 155-202. Print.

Bissel, Tom. "Whimsy in the Face of Terror." Review of *Extremely Loud and Incredibly Close*, by Jonathan Safran Foer. The New Leader Mar./Apr. 2005: 28-29. MasterFILE Premier. Web. 18 Aug. 2012.

Carlini, Matthew Francis. "The 'Ruins of the Future': Counter-Narratives to Terrorism in the 9/11 Literature of Don DeLillo, Jonathan Safran Foer, and Ian McEwan." MA thesis. U of Tennessee, 2009. Web. 14 Nov. 2011.

Couch, Natalie Eva. "Pictures, Puzzles, and Missing Pieces: The Childlike Solution to Trauma in the Mature Novel." MA thesis. Clemson U, 2008. MAI 46.5 (2008): item AA1450718. ProQuest Dissertations and Theses. Web. 6 Nov. 2011.

"Discussing Metamodernism." Notes on Metamodernism. 26 Mar. 2012. Web. 24 June 2012.

Fessler, Nadine. "Illuminating Language." Notes on Metamodernism. 29 Sept. 2011. Web. 19 Dec. 2011.

—. "To engage in literature." Notes on Metamodernism. 29 May 2012. Web. 11 June 2012.

Foer, Jonathan Safran. "About the Typefaces Not Used in This Edition." *The Paris Review* 42.156 (2000): 383-89. LION. Web. 3 Nov. 2011.

—. "A Primer for the Punctuation of Heart Disease." *The Unabridged Pocketbook of Lightning*. London: Penguin, 2005. Print. 1-9.

—. *Extremely Loud and Incredibly Close*. New York: Houghton, 2005. Print.

—. Introduction "What We Say We Are." *The Diary of Petr Ginz*. Ed. Chava Pressburge. Trans. Elena Lappin. New York: Grove Press, 2007. vii-x. Print.

—. "Ma Jian Introduced by Jonathan Safran Foer." *Words Without Border: The World Through the Eyes of Writers*. Ed. Samantha Schnee, Alane Salierno Mason, and Dedi Felman. New York: Anchor Books, 2007. Print. 3.

Hartness, Paula B. "Po Pomo: The Post Postmodern Condition." MA thesis. Georgetown U, 2009. MAI 47.5 (2009): item AAT 1464741. ProQuest Dissertations and Theses. Web. 21 Nov. 2011.

McAdams, James. "David Foster Wallace's 'Gfhrytytu': The Opening of Another Ethics." MA Thesis. Villanova U, 2011. MAI 50.1 (2012): item AAT 1497130. ProQuest Dissertations and Theses. Web. 16 Nov. 2011.

McLaffery, Larry. "An Interview with David Foster Wallace." *Review of Contemporary Fiction* 13.2 (1993): 127-50. Academic Search Complete. Web. 18 Aug. 2012.

McLaughlin, Robert L. "Post-Postmodern Discontent: Contemporary Fiction and the Social World." *symplokē* 12.1/2 (2004): 53-68. JSTOR. Web. 16 Nov. 2011.

Schoop, Katelynn. "'Hearts Ripe, Brains Aglow': Narrative, Subjectivity, and Feeling in Post-postmodernism.'" MA thesis. Trent U., 2009. MAI 48.2 (2010): item AAT MR53248. ProQuest Dissertations and Theses. Web. 16 Nov. 2011.

Siegel, Harry. "Extremely Cloying & Incredibly False." Review of *Extremely Loud and Incredibly Close*, by Jonathan Safran Foer. New York Press 20 Apr. 2005: n. pag. Web. 22 Aug. 2011.

Smith, Rachel Greenwald. "Organic Shrapnel: Affect and Aesthetics in September 11 Fiction." *American Literature* 83.1 (2011): 153-74. Academic Search Complete. Web. 26 Nov. 2011.

Timmer, Nicoline. *Do You Feel It Too?: The Post-Postmodern Syndrome in American Fiction at the Turn of the Millennium*. Amsterdam: Rodopi, 2010. Print.

Tyson, Lois. *Critical Theory Today A User-Friendly Guide*. New York: Garland Publishing, Inc., 1999. Print.

Vermeulen, Timotheus and Robin van den Akker. "Notes on metamodernism." Journal of Aesthetics & Culture [Online], 2 (2010). Web. 24 June 2012.

Versluys, Kristiaan. "A Rose Is Not a Rose Is Not a Rose: History and Language in Jonathan Safran Foer's *Extremely Loud and Incredibly Close*." *Out of the Blue: September 11 and the Novel*. New York: Columbia UP, 2009. 79-119. Gale Literary Database Contemporary Literary Criticism. Web. 30 Sep. 2011.

Wallace, David Foster. "Brief Interviews with Hideous Men." *Brief Interviews With Hideous Men*. New York: Little, Brown, 2007. Print.

CHAPTER FIVE

SUCKER PUNCH: A CARNIVALESQUE STEAMPUNK FAIRY TALE

KRISTEN BRADLEY

Sucker Punch is a movie released in 2011 and written and directed by Zack Snyder, most recently known for his work on *300* and *The Watchmen*. Snyder's visual aesthetic is predominant within the film, combining stunning graphics and CGI with over-blown fight scenes. His dark, dreary reality of harsh greys and blacks soon melts away into a colorful fantasy laden with undead armies and mythological creatures. Snyder may be known for his striking visuals, but this is the first time he has written a film not based on preexisting source materials.[1] As such, Snyder was able to utilize a number of preexisting genres, including the fairy tale. Adding his own contemporary twist and modern sensibilities, Snyder creates a modern-day fairy tale containing elements of the carnivalesque and reflective nostalgia.

The movie revolves around Baby Doll, a wide-eyed blonde played by *A Series of Unfortunate Events*'s Emily Browning. It is set in three levels, which for the purposes of this essay will be called the three realities. In the first reality, following the death of her mother, Baby Doll's stepfather sends her to a mental institution. Through shady dealings with a custodian at the asylum, the stepfather arranges for a lobotomy to be performed on Baby Doll. The onset of this procedure marks a shift out of the extradiegetic narrative into the first hypodiegetic narrative, or the second reality.[2] In this reality, the asylum is transformed into a night club/brothel.

---

[1] Both *300* and *The Watchmen* were based on comic books: the former by Frank Miller and the latter by Alan Moore (Corliss).

[2] The term extradiegetic narrative refers to an outer or matrix narrative which contains another, inner narrative, often referred to as a hyponarrative or a hypodiegetic narrative. For more information on these terms, see Shlomith Rimmon-Kenan's seminal work *Narrative Fiction*.

The on-goings of this reality represent Baby Doll's reimagining of her first five days at the asylum, which takes place entirely within her mind and within the second between the doctor placing a pick over her eye and then driving it in to complete the lobotomy. During this reimagining, Baby Doll is launched into yet another hyponarrative, or the third reality. Music and dancing act as the trigger which marks the shift in realities and as a framing device for these scenes.

The story centers around five girls: Baby Doll, Sweetpea, Rocket, Amber, and Blondie, which are patients at the mental institution in the first reality and "dancers" at the brothel in the second. The asylum workers are also transformed in this reality. Dr. Gorsky, the head psychiatrist, becomes Madam Gorsky, the dance instructor, while Blue, the custodian, becomes the club owner and mob boss. In Blue's club, the girls "are the main attraction." Desperate to escape the tyrannical rule of Blue and the prying eyes and hands of the older men who frequent the club, Baby Doll creates a third reality, a fantasy realm she enters when forced to "perform" at the brothel. Here she meets the Wise Man, who assists her in developing a plan to escape. To do so, she must collect five items: a map, fire, a knife, a key, and a secret item which, according to the Wise Man, will involve "a deep sacrifice and a perfect victory."

The film's setting provides an interesting dimension to the story. The year is rather obscure. The film takes place in the past, like many fairy tales, but not so distant a past that it is un-relatable to modern audiences. It is unmarked time, lending itself to the typical fairy tale "once upon a time." The director states the film is set in the 1960s, but the plot line seems more fitting for the 1940s when lobotomies were more prevalent (Sharkey).[3] Clues to the story's actual timeline are infrequent, and the movie is described by Sharkey as a "violent mash-up of Dickensian dark morality with Moulin Rouge couture." The costuming is partially Victorian, partially anime, with one critic calling Baby Doll's outfit a "Japanese schoolgirl outfit as retailored by Victoria's Secret" (Corliss). The multiple realities of the narrative are populated by folk tale characters: "soldiers, peasants, animals, and supernatural creatures (witches, fairies, elves, dwarves, goblins, giants)" (Zipes, *Fairy Tales* 7). Though fairy tales rarely depict "members of the bourgeoisie or the church," or machines or other "signs of industrialization" (7), the blurring of time periods utilized by the movie lends itself to the addition of these characteristics through the use of the carnivalesque and reflective nostalgia.

---

[3] The Nobel Prize for Physiology or Medicine was awarded for the procedure in 1949 despite the fact that the procedure was quickly falling out of favor (Boeree).

The carnivalesque is often defined as a "time out of time." It is marked by grotesque body imagery, often through sexuality, and the symbolic inversion of hierarchy (Babcock, Bakhtin). While the carnivalesque "time out of time" occurs in reality as a break from everyday life, for Baby Doll it occurs within her fantastic second and third realities. The second reality is quite literally a "time out of time" in that the entire five days take place within only a second. The third reality is a similarly literal interpretation, this time taking Baby Doll out of her everyday activity and into a fantastical realm. Both of these periods are marked by numerous manifestations of the carnivalesque, most significantly through the form of hierarchical inversion in the second reality. The psychiatrist, Dr. Gorsky, is transformed into a dance instructor, subservient to the brothel owner, Blue, a mere custodian in the first reality. Further symbolic inversion of hierarchy is also included in this level of reality. The most often parodied entities in the carnivalesque are figures of authority: government/monarchy, religious figures, and soldiers/military (Babcock). While Baby Doll is dropped off at the asylum by her stepfather in the first reality, in the second, he is transformed into a perverse priest, made more grotesque by the implied sexual abuse Baby Doll suffered at the hands of her stepfather. The city's mayor is also a frequent client of the brothel, even having a regular girl, Amber. The mayor is portrayed as corrupt and greedy, surrounded in lavishness. This excess is sardonically parodied, presented in the extreme, and his position is mocked by the sheer extremity. He is described as always having "a big stogie in his mouth," and he continually blows smoke rings. Viewing the stogie in terms of the carnivalesque, it is a phallic symbol, which subverts the mayor's hierarchical status when the end falls off, symbolically emasculating him. It is particularly poignant that this scene occurs within the second reality, Baby Doll's fantasy. When her real life, first reality, is marked by male sexual aggression, her carnivalesque inversion would have to include the emasculation of a male figure and the subversion of this aggression, impotence. Similarly, the third reality includes epic battle scenes against undead German World War I soldiers and hyper stylized Japanese feudal warriors. One of these feudal warriors dons a traditional wire carnival mask complete with phallic nose. This warrior is defeated in battle by Baby Doll. All five girls battle the creatures in an act of empowerment, using large, masculine guns and weapons. Amber is given control of the massive vehicles, including a plane, helicopter, and a giant robot equipped with firearms. Amber always assumes this masculine role with a lollipop in her mouth, drawing an interesting parallel through oral fixation between Amber and her exclusive customer, the mayor.

Aside from the carnivalesque "time out of time," the fairy tale "once upon a time" is also created through the use of reflective nostalgia. According to Svetlana Boym, reflective nostalgia is the "complex folding of time" (qtd in Dawdy). It is a combination of "retro and eclectic interior design styles, which are guided not by a desire to accurately reproduce past place-times, as would a period room of a museum, but rather by an impulse to play with future possibilities by jumbling anachronistic furniture and artifacts" (Dawdy). Rather than only utilizing ideas and artifacts from a single, particular time period, reflective nostalgia indiscriminately blends artifacts from multiple eras. This idea takes many forms within popular culture, the most prevalent of which is steampunk. Steampunk is characterized by a blending of romantic Victorian costuming and ideals of the steam era with modern technologies. According to Rebecca Onion, steampunk draws from "such diverse sources of influence as Dickens, future science fiction, cyborg theory, and modernist art" (142). It is generally limited, however, in its use of the past, focusing mainly on the nineteenth century (Onion).

Several tropes of the stempunk genre are present within the third reality in *Sucker Punch*. These include, most notably, the steam and clock-powered World War I-era German soldiers. According to the Wise Man, German engineers discovered a way to reanimate their fallen soldiers. These undead soldiers are reflective of the steampunk ideal of utilizing "robot technology, but executed with nineteenth-century materials" (Onion 141). The battle against these soldiers culminates in an extravagant chase involving a giant robot operated by Amber and the Germans' zeppelin. According to Onion, the zeppelin is the "prototypical example of steampunk interest in disastrous technology" (150). This fascination often culminates in the replication of the "spectacular" explosion of the *Hindenberg* in 1937 (Onion). Both the ship and the pursuing giant robot are massive even by modern movie standards, creating another standard steampunk trope: the "moment in which characters crane their necks to take in the full expanse of the technological object they are confronted with" (154).

Because many limit the scope of steampunk to the utilization of nineteenth century fashion and ideals, we must look elsewhere to explain alternate creative anachronisms. For this, we must use the broader term "reflective nostalgia," which anthropologist Shannon Lee Dawdy describes as a "temporal folding in popular culture." This collapsing of time is typically demonstrated through creative juxtaposition of artifacts, but the idea is certainly reminiscent of the collapse of time periods used for the film's setting. Dawdy further states that reflective nostalgia is known for "purposefully entangling moments of its progressive timeline." Snyder's film mixes moments of the 1940s, the 1960s, Victorian era, steam technology, fantasy, and futuristic technologies. Many of the fight scenes in the third reality follow the same idea. The most prominent example occurs when the girls fly a massive helicopter, equipped with military grade missiles, and fall in pursuit of an old-time train engine. Once the girls board the train, they are faced with futuristic cyborgs. A long pan shot shows the train speeding towards an alien city. While these urban, technological elements are not typical of the fairy tale, the temporal blending caused by reflective nostalgia makes it impossible to place the tale in a specific time period, lending to the fairy tale trope of "once upon a time."

The plot of the movie also contains the characteristics and functions of the typical fairy tale. On a basic level, almost none of the characters have real names. This is an important quality, especially considering the fairy tale's origins in oral tradition. Jack Zipes writes in *Why Fairy Tales Stick* that "names are rarely used in a folk tale" (49). We see this in *Sucker Punch* as the five girls are referred to only by their brothel nicknames; we are never given their real names. Likewise, other important characters, including Baby Doll's stepfather and sister, the cook and the Wise Man are never given names at all. The same applies to the second reality's villains, the Mayor and the High Roller. As Zipes states, these "characters function according to their status within a family, social class, or profession" (*Why* 49). As such, they are referred to only according to their respective roles rather than by true names. The only characters that prove exceptions to this rule are Blue, the custodian/brothel owner (if that is even his real name), and Madam/Dr. Gorsky. Oddly enough, these are the two most featured adult characters in the film. In this context, the inclusion of their names helps to establish their authority over the nameless girls in the brothel.

Though *Sucker Punch* is a movie about girls working in a brothel, it is strangely desexualized. Movie critic Richard Corliss writes, "Odd, then, that the film's erotic temperature is so tepid." This, too, is owing to the fairy tale nature of the story. The Grimms' fairy tales went through a total of seven revisions between their initial publication in 1810 and 1857. According to Zipes, "Wilhelm Grimm, the more conservative of the two brothers, did most of the revisions, and it is commonly known that he endeavored to clean up the tales and make them more respectable for bourgeois children" (*Fairy Tales* 48). This "cleaning up" actually entailed "sanitiz[ing] [the tales] to the point where the erotic element disappears" (52). This is strangely reminiscent of Corliss's complaint about the film.

Zipes's prime example of sanitization is the tale of the Frog Prince. In the 1810 version of the tale, the Frog Prince tells the princess that he will retrieve her ball only if she will sleep with him. As Zipes states, "There is no beating around the bush . . . It is explicitly sexual" (50). Later versions of the story do not include this element. Like the later versions of the Frog Prince, *Sucker Punch* has been stripped of its explicit sexual elements. Surely plenty is implied, but these must be inferred from the tale. When Baby Doll's stepfather schedules the lobotomy, he does so to prevent her from being able to tell the police what he did to her. The implied sexual abuse is never shown, nor stated explicitly, but it is clear what is insinuated. This is made more explicit when Blondie states matter-of-factly that "She's no virgin," in reference to Baby Doll. Further, it is only implied that the nightclub is a brothel, when Rocket lies down on a spinning, red velvet bed and states, "We are the main attraction." For the majority of the second fantasy, however, the girls are portrayed merely as erotic dancers, and the madam becomes a dance instructor. There are vague references to the activities of a brothel, but most attention is paid to the dance elements. In keeping with the Grimms' revisions, however, this too is strangely desexualized. The audience never sees the girls' erotic dancing. It is implied that Baby Doll is an especially talented dancer, but every time she begins to do so, her mind takes her into the third reality, implying that this reality is her way of coping with the horrible display she is forced to perform. Further, threats of rape are never fulfilled or even fully stated. Certainly there are sexual overtones to Blue's threats, even to Madam Gorsky, but he always stops short of saying anything explicitly sexual. When threatening Baby Doll, Blue states, "if I weren't about to make a small fortune on you with the High Roller, I would . . ." This sentence is left incomplete.

In typical fairy tale fashion, the hero of the story must go on a quest or an adventure. This is a combination of ideas from Vladimir Propp, who identified thirty-one functions which comprise the structure of the fairy tale, and Max Luthi.[4] In Propp's ideas the hero must first realize a lack and then go "in search for aid (intermediaries) to achieve happiness" (Zipes, *Fairy Tales* 4). According to Luthi, the hero is a "wanderer charged with carrying out a task" (4). Both are present in the film. Upon Baby Doll's first entrance into the third reality, she realizes her lack and meets her intermediary figure, the Wise Man (Propp 9). The Wise Man must cause

[4] In Propp's model, functions are "the fundamental and constant components of a tale that are the acts of a character and necessary for driving the action forward" (Zipes, *Why* 49).

Baby Doll to realize her lack (Propp 8a). He asks, "What do you want?" Baby Doll does not know, but he asks her again. She realizes suddenly, stating, "a way out." She wants to escape from the asylum; she wants freedom. This quest for freedom then becomes the task with which she is assigned, per Luthi. The composition of the folk tale and likewise *Sucker Punch* "is a detailing of the ways in which the hero takes steps to survive and complete his [or her] mission" (Zipes, *Fairy Tales* 5).

Along this journey, the hero will meet both friends and foes. According to Zipes, "the friend is a mysterious and supernatural individual or creature, who has unique powers and gives the protagonist gifts. Sometimes there are at first three different animals or creatures that test the protagonist to see whether he is worthy of their help" (*Why* 49-50). Utilizing Propp, the friend acts as "mediation" a figure who guides the hero. He is important in multiple functions: the first function of the donor (Propp 12), the test described by Zipes above; the hero's reaction (Propp 13), presumably overcoming the test, and provision or receipt of gift (Propp 14) and spatial transference or guidance (Propp 15). Each of these ideas is present within the movie. The friend or intermediary in *Sucker Punch* is the Wise Man. The Wise Man acts as Baby Doll's guide through the third reality, but first he must test her worth. After telling her how she can accomplish her goal of escape, he states, "Oh yeah, and one more thing. Defend yourself." Suddenly she is faced with three feudal Japanese warriors and must indeed defend herself. Only after completing this test does the Wise Man help her to locate the five items she needs to free herself. To do this, he provides her with both guidance and gifts along the way, including weapons and means of transportation.

The villains in *Sucker Punch* are present in multiple levels of reality. Zipes defines villains as being "those who use words and power intentionally to exploit, control, transfix, incarcerate, and destroy for their own benefit" (*Why* 51). In Propp, these villains also serve a number of functions. First, the villain takes action against the hero (reconnaissance, Propp 4). In the first reality, this villain is the stepfather. We have already discussed his implied rape or sexual abuse of Baby Doll. This would certainly qualify as negative action against the hero. The next function is delivery (Propp 5), wherein the villain receives information about the victim. In this case, the stepfather reads the mother's will leaving all of her possessions to her two daughters. The stepfather reacts negatively, causing harm to a member of the family (villainy, Propp 8). He once again comes after Baby Doll, but, unable to get her, instead attacks the little sister, eventually leading to her death. A final function of the villain is trickery (Propp 6), and in this first reality, we see Baby Doll's stepfather tricking the authorities into committing her to an institution. He states, "I told the police she went crazy after her mother died. The truth is actually a bit more complicated than that." The villain again appears, though in varied forms in the second and third realities of the movie. Arguably, once Baby Doll enters the asylum, Blue becomes the main villain, both in the first and second realities[5]. More wildly abstracted villains appear in the third realities, often as allegorical representations of villains in the second reality. When Sweetpea is tasked with stealing a map from Blue's office, the third reality depicts a reconnaissance mission to recover a map from the trenches of undead German soldiers. Likewise, when Amber is sent to steal the mayor's lighter in the second reality, the girls face off against an army of orcs to steal two fire-making crystals from a dragon. Just as Amber is nearly caught by the mother dragon in this reality, she nearly caught by the Mayor in the second. Given the earlier description of the mayor, it is especially interesting that his allegorical counterpart is the dragon, a mythological creature often portrayed as greedy. The final mission to steal a knife from the cook in the second reality ends badly on both realities. As Rocket dives between her sister and the cook's knife in the second reality, her third reality self is the one who pushes her sister off a train just before it explodes, sacrificing herself for the well-being of the other girls. This is a poignant scene which beautifully juxtaposes the on-goings of the two realities, cleverly using the frame device (music) to trigger breaks between the two worlds.

The hero's journey is often marked by a particular set of functions, as well as a number of epic laws as proposed by Axel Olrik. One function that plays a predominant role within the film is that which Propp calls interdiction (Propp 2), typically followed by violation (Propp 3). First, the hero is warned to not do a particular thing (interdiction) and then ignores this warning (violation). This often occurs both on the large scale and the small scale within a single story. The large scale interdiction and subsequent violation makes the girls' epic journey possible. Sweetpea often warns Baby Doll to not try to escape. "No one's ever escaped from here," she says, adding that the last girls who tried were killed. Baby Doll, of course, tries anyway. On a smaller scale, mini-adventures within the tale are often marked by interdiction. Before the girls go after the two crystals that produce fire, which happen to be inside the throat of a baby

[5] Not only does Blue arrange her lobotomy, he forges the doctor's signature on the authorization, and there is again implied sexual abuse ongoing between himself and the patients. If the girls in the asylum were being sexually abused within the institution, this could explain Baby Doll's reimagining (the brothel) as a coping mechanism.

dragon, the Wise Man states, "Oh yeah, and one last thing. Don't wake the mother." This, like most interdictions, is followed by violation. The girls do wake the mother. The violation nearly causes their downfall, though the girls do get away with the fire.

The near downfall in this case is a smaller version of what Zipes calls peripety — a "sudden fall in the protagonist's fortunes, which is only a temporary setback" (*Why* 50). A more serious version of this setback occurs in the third group battle (the battle against the cook) and its immediate aftermath. As stated above, the cook stabs Rocket when she places herself between him and her sister, Sweetpea. This setback in the girls' plan is made even more tragic when Blue catches onto their plan, identifying Baby Doll as the ring leader by the list of objects on her blackboard (branding, marking, Propp 17). As viewers, we find out that Blondie has told Madam Gorsky and Blue about the plan. Blue locks Sweetpea away in a closet, then makes an example of Amber by shooting her point blank. He then tells Blondie, "We don't like snitches," next turning his gun on her.[6]

Here we are shown the important use of Axel Olrik's epic laws. Olrik believes the tale is marked by the presence of a number of characteristics that recur in many (if not all) folk or fairy tales. His first law, the Law of Opening and Closing, denotes the presence of a narrator, which fittingly both opens and closes the movie *Sucker Punch*. Further, the tale is marked with the Law of Repetition. This is typical of oral tales, but literary adaptations and film adopt it in the form of repeated lines. One such line in *Sucker Punch* is the Wise Man's refrain "Oh yeah, and one more thing" or "Oh yeah, and one last thing," repeated throughout the course of the movie. The number five seems to be an important theme in the movie,[7] but Olrik's Law of Threes is no less present. The movie is set in three realities. There are also three security guards employed at the asylum, as well as three hierarchical figures: the cook, Madam Gorsky, and Blue. In Baby Doll's test battle, she fights three feudal Japanese warriors, and the girls' big adventure begins three days before the High Roller comes. They collect three items through three epic fantasy battles in the third reality before three of the girls are killed. During this time, Baby Doll "dances" only three times. Olrik's final law, the Importance of Initial and Final Position, however, is particularly important in this film and requires further explanation.

---

[6] If we see the girls as a constructed family unit, this again fulfills Propp's function, villainy (8), or enacting harm against a family member.

[7] There are five girls, five items, and five days until the arrival of the High Roller.

The film opens with Baby Doll sitting in her room at home in the first reality. We find out that her mother has just passed away (Propp 1, absentatia). This is what Olrik would call the initial position. The principle character of the film always comes first. The expected adherence to this idea leads to an unexpected twist near the end of the movie. Following the girls' peripety, or temporary setback, they must eventually succeed (at least for this story to be called a fairy tale), and the villains must be defeated (Propp 18). After Blue has killed Baby Doll's friends and threatened her with implied sexual overtones, Baby Doll stabs Blue in the neck with the knife[8] stolen from the chef. She then steals the master key from Blue's necklace. Having defeated the villain, Baby Dolls rescues Sweetpea and the two girls utilize the items to enact their plan. Victory seems imminent until the girls spot a gang of men standing in front of the asylum/nightclub. Baby Doll then realizes that she (her head) is the fifth item, the "great sacrifice" to which the Wise Man had referred. She must sacrifice herself to allow Sweetpea to go free.

For the story to be a fairy tale, the protagonist must have her "happily ever after." The movie *Sucker Punch* begs the viewer to ask the question "whose story is this?" As Baby Doll tells Sweetpea, "This was never my story. It was yours." Baby Doll distracts the men, enabling Sweetpea's escape. Baby Doll is captured, and we are suddenly plummeted back into the first reality as the doctor's pick enters Babydoll's brain. The movie does not end there, however. Sweetpea has reluctantly throughout the story realized her lack; initially and continually, she objected to the plan of escape. In the end, however, only for her is the lack liquidated (Propp 19). Sweetpea is able to return home to her parents, while Babydoll becomes Sweetpea's intermediary (9). She offers Sweetpea guidance and advice, stating, "Stay off the roads. Don't screw this up," though she does not replace the Wise Man. The Wise Man appears once again to help when Sweetpea is pursued (Propp 21) by the police investigating the escape.[9] The final functions are left to the viewer's imagination as Sweetpea embarks on her journey home. The Wise Man emphasizes this, stating, "We have a long way to go." Leaving the viewer to speculate as to Sweetpea's happily ever after, *Sucker Punch* ends. The last shot focuses exclusively on Sweetpea, the story's unlikely heroine, in what Olrik would call the final position. According to Olrik, the final position is reserved for the character that arouses the audience's sympathy. We have seen her

[8] While this occurs in the second reality, it is implied that it also occurs in the first reality, when Dr. Gorsky states that Baby Doll has "stabbed an orderly."

[9] It is unclear whether this occurs in the first or second reality, or both, as Baby Doll is cited as having helped a fellow patient escape in both realities.

plight, and we can now feel hope that she will finally make it home. Zipes cites this as the main function of the fairy tale; "At the center of attraction is the survival of the protagonist under difficult conditions, and the tales evoke wonder and admiration for oppressed characters no matter who they may be" (*Why* 50). This "hope for change" is key to the fairy tale's ending (50).

Sucker Punch is an unexpected fairy tale. It comes to us in the form of a high budget film, which did poorly in the box office and garnered numerous critical reviews, mostly negative. The film's flashy CGI and video game fight scenes are used to mask what many feel to be a substandard story line. Upon closer review, however, it is apparent that the film utilizes archetypal storytelling elements, many of which are centuries, even millennia, old. The combination of these elements with contemporary sensibilities, reflective nostalgia and steampunk elements, and the tropes of the carnivalesque creates a unique fairy tale that adds depth and character to Snyder's best film.

References

Babcock, Barbara A. "Introduction." In *The Reversible World,* ed.
Barbara A. Babcock, pp. 13-38. Ithaca and London: Cornell University Press, 1978. Print.

Bakhtin, M.M. "Introduction." In *Rabelais and His World*, pp. 1-58. Bloomington: Indiana University Press, 1984. Print.

Boeree, C. George. "A Brief History of the Lobotomy." 2001. Web. 10 April 2011. http://webspace.ship.edu/cgboer/lobotomy.html

Corliss, Richard. "*Sucker Punch* Review: Don't Drink the Kool-Aid." *Time*. 25 March 2011. Web. 10 April 2011.

Dawdy, Shannon Lee. "Clockpunk Anthropology and the Ruins of Modernity." *Current Anthropology* 51.6 (2010), n.p. Web. JSTOR. 4 April 2011.

Olrik, Axel. "Epic Laws of Folk Narrative." Ed. Alan Dundes. *International Folkloristics*. New York: Rowman and Littlefield, 1999. 83-98. Print.

Onion, Rebecca. "Reclaiming the Machine: An Introductory Look at Steampunk in Everyday Practice." *Neo-Victorian Studies*. 1.1 (2008), 138-163. Web. 4 April 2011. http://www.neovictorianstudies.com

Propp, Vladimir. "The Structure of Russian Fairy Tales." Ed. Alan Dundes. *International Folkloristics*. New York: Rowman and Littlefield, 1999. 119-130. Print.

Rimmon-Kenan, Shlomith. *Narrative Fiction: Contemporary Poetics*. New York: Routledge, 2002. Print.

Sharkey, Betsy. "Movie Review: 'Sucker Punch'." *Los Angeles Times*. 25 March 2011. Web.10 April 2011.

Snyder, Zack, dir. *Sucker Punch*. Warner Bros., 2011. Film.

Zipes, Jack. *Fairy Tales and the Art of Subversion*. New York: Wildman Press, 1983. Print.

—. *Why Fairy Tales Stick: The Evolution and Relevance of a Genre*. New York: Routledge, 2006. Print.

PART III:

IMAGINATION IN THE DIGITAL AGE: CLASSROOM PRODUCTION STRATEGIES

CHAPTER SIX

THE GRAPHIC NOVEL: TEACHING AND PRODUCING TEXTS

MARLY ROBERTSON

I arrived to class one afternoon with a box full of shrink-wrapped booklets with clean lines and muted colors on the front covers. The moment I walked through the door, I could feel a heightened energy in the room. My students asked if the box contained copies of their comic anthologies, and indeed it did. As the minutes ticked on and the last students strolled into the classroom, the students who had arrived early had clearly become antsy. "Why do we have to wait?" they asked. I was waiting for the big reveal; I wanted everyone to be present before handing out the comic book collections. After the final students entered the classroom, I began passing out the shrink-wrapped packages: collections of comics that my students had created. I could hear crinkling plastic as my students began ripping the packages open. There were gasps and squeals and gigantic, toothy grins as the pages rustled and each student found his or her comic in the book. Moments after the students had seen their own comics, they looked over the others and complimented their peers on a job well done. This was one of the most rewarding moments of my teaching career. My students had created a text that brought them together as friends and creators instead of merely as fellow students.

During the fall 2011 semester, I taught English 210: The Graphic Novel at Nicholls State University. I planned to demonstrate to my students and fellow faculty members that comics are not "crude, poorly drawn, semiliterate, cheap, disposable, kiddie fare;" comics are literature, albeit a non-traditional form of literature (McCloud 3). I hoped that students, whether they were comics fans or not, would see that comics were not just "fun reading;" they could have much deeper meaning and could be just as complex as any classic literature.

With such a non-traditional course, I also had to decide if it was appropriate to ask my students to complete "traditional" essays. Ultimately,

I concluded that students should create their own comics in order to showcase the knowledge they had acquired throughout the course of the semester. The following essay demonstrates the struggles and successes of their non-traditional writing assignment, a comic that was then bound into an anthology of all the students' works.

The assignment itself, which was adapted from University of New Orleans instructor Beth Blankenship's assignment, was a somewhat last minute decision made after discussing the success Beth had in the past with this assignment. The original goal was for students to demonstrate the various comic elements we had discussed in class, such as paneling, closure, effective utilization of abstract images, and the intersection of the verbal and visual. I also wanted students to demonstrate more basic elements of narrative like plot and character development. I had no idea at the time that the project would foster community in the classroom to the degree that it did.

The assignment asked students to work either individually or in groups to create a comic 3-5 pages in length, with several panels per page; they were permitted to create longer comics if it did not interfere with their ability to complete the comics by the due date (Blankenship n. pag.). The students were asked to create an original story or to retell an old story with a twist or from a new perspective. The assignment outlined several benchmarks for the students to keep them on task for the duration of the project (about four weeks). They were required to submit a proposal, a working script, a list of characters with sketches, a storyboard, a rough draft, a final draft and several journal entries describing their progress (Blankenship n. pag.).

Upon being presented with the project, the students were remarkably receptive and were more convinced than I was that this project would be successful. The vote to do the creative assignment instead of a traditional paper was nearly unanimous, and those who were hesitant were persuaded when they realized they did not have to be brilliant artists in order to be successful. The assignment encouraged the students to use clip art or found images if they were not great artists (Blankenship n. pag.). Several students did this, but several sets of students paired up so that one person would create the narrative while the other did the art work. Still others decided to own their lack of artistic talent and chose to create stick figures with which to tell their stories. What is perhaps the most remarkable discovery here is that some of the comics drawn with stick figures were just as successful as those that were perfectly and painstakingly drawn even though they were less aesthetically pleasing. This is largely because the stick figures were able to represent purpose and movement in spite of their basic, unrefined form, perfectly illustrating McCloud's discussion of icons yielding amplification of meaning through simplification of images (30).

When I allowed students class time in which to work on their projects, the room was abuzz with ideas and excitement. This is, as most teachers know, not always the case for other types of group assignments. All too often students are simply plugging along to get the work done--waiting for the clock to indicate that class is over. In other cases, students do not do any work at all because they prefer to let the other students do the work for them. Frequently, in peer review workshops, students do not truly value what they are doing nor do they value the opportunity to collaborate with their peers. This project was entirely different since the students were invested in their projects and taking pride in them. On one in-class work day, a few students, when they had finished their own work, began offering artistic and narrative suggestions to peers in other groups. They all seemed very excited about the work their peers were doing and could not wait for the next opportunity to work together. This experience for me, and it seems for the students as well, was truly a remarkable one because everyone was working together to create the best work possible.

The next phase of this creative project was print publication, which created an array of obstacles. I quickly discovered that I was extremely naïve about the costs of print production. In order to bind an entire collection, I was quoted prices around $400 for roughly 40 full-color, two-sided pages. At a small, regional university that is particularly low on funds, this price was simply out of the question. I nearly gave up on the project, and when I told my students about the funding issue, many of them offered their own money to complete the project. This was yet another heartening aspect of this assignment. How often do students in a community that is not affluent offer their money for educational purposes? In spite of their generosity, I just could not bring myself to take my students' money. There had to be another way. I considered creating a blog or website as a forum for the students to present and share their work, but that just would not be good enough. My students and I all wanted a tangible product and they were genuinely upset when they thought they might not get one. After asking around our department, I was granted permission to use our brand new scanner and large-scale color printer, which had been provided via Student Government funds. Conveniently, we also had the correct paper size in stock. I was elated.

But without the money to have our book printed professionally, I now had a colossal amount of work on my hands. I had originally requested electronic copies or scans of the comics, but not surprisingly, most of my students did not have access to scanners. So my first task was scanning each page. There were several issues with the quality of the scanned images. Some images were not dark enough or were too dark, but there was little that could be done about that on such a compressed schedule. I was able to make some slight alterations electronically, but some images were just poorly rendered. I shudder to think how many hours I spent on the task of scanning images; it definitely siphoned several days' worth of office hours and spare time on campus.

The next part of production was formatting and printing. I essentially merged the PDF files into an order I was happy with since there was no time to obtain student input on this matter. I decided to bookend the collection with two of the strongest pieces. Then, I had to tinker with the page layout in order to print the book onto 14x17 inch pages. I quickly discovered that printing an entire book at a time took hours, and it was much faster to print each page and collate them myself. I spent two days setting the printer up in between classes and scurrying between the media projects room and my own classes in order to get everything done before the last day of class. Finally, I sent my collated pages to a local print shop for stapling, folding, trimming, and shrink wrapping. This cost me a total of ten dollars, a far more manageable price than expected.

The assessment of this project provided another set of obstacles. I am not accustomed to grading this type of creative assignment so I found it very difficult to assign grades to something that was completely subjective. The assigned benchmarks came in handy here as each benchmark provided a concrete grade based on completion and following certain rules. The benchmarks were the bulk of the final grade (about 70%) and the completed comic made up the other 30% of the students' grades. It was difficult not to focus on the aesthetics of each comic as that was not the ultimate goal. I often had to step back from my initial reaction to the comic and remind myself that students were allowed to draw stick figures. Having said that, it was hard not to automatically reduce a student's grade for having a less-than-beautiful final product. I ultimately had to focus on how well the students followed comic customs and expectations. I examined the comics for quality narratives that "flowed" effectively. I examined the use of paneling in each comic and how the students used the gutter to their advantage. Additionally, I took into consideration the use of language and images and their effectiveness. One of my students chose to tell his entire story with a narrator rather than using dialogue. This student also valued images over words throughout the comic. Interestingly, this was one of the most effective comics. Other students relied on language quite heavily and their images had fewer details. This led to a great deal of variety in the students' work, but there was not one type of comic that was inherently more effective than another, which I found compelling.

There were, of course, some problems associated with group assessment. What does an instructor do when some students are not pulling their weight in their groups? While it was often obvious who was really doing work during our in-class sessions, I had to rely on students to keep me abreast of what was going on outside of the classroom. I was surprised by how willing my students were to confess that their peers were not doing their work. I took this as a sign that my students truly valued the work they were doing and that they were invested in the final product. It seemed as if the students were taking this project more seriously than they would have taken a traditional writing assignment; they were genuinely proud of their work and did not want anyone else negatively affecting it. The students seemed offended when their peers were not willing to put forth the same amount of effort that they were. I was very careful not to penalize the students in the group who were working hard and doing their assignment correctly and appropriately. Fortunately, there were very few students who were affected by this situation.

Given the opportunity to teach this assignment again, there are very few changes I would make and most are associated with process work. First, I would give the students the entire second half of the semester instead of just four weeks to work on their comics. The project seemed very rushed in four weeks, and the students had a difficult time keeping up with each of the benchmarks. I would also merge the storyboard and rough draft of the comic. Many students did not understand how to make an effective storyboard, and several just turned in a rough draft for that assignment. Additionally, since so many students were drawing by hand, asking for a rough draft meant an excessive amount of work for the students, and there simply was not enough time to make such a request fair to them. While I tried to convey to them that their rough drafts could be extremely rough, consisting largely of sketches, most of them turned in polished work anyway; this is yet another example of their enthusiasm for this project. Finally, I would allow students several more opportunities for group work in class. They only worked in class on three occasions and that simply was not enough. It would be beneficial for them to work together, batting around ideas instead of just delegating tasks and working individually. The additional in-class work days would allow more opportunities to collaborate with one another on a unified project.

There were also several successes associated with this project. My students' enthusiasm led many of them to create comics that were at least double the minimum page length. How often does that happen with a traditional academic essay? They were also extremely eager to share their work with others and to complement each other on their projects. Many of the students' concepts were also fascinating and unique. For instance, one of my students chose to re-tell *Little Red Riding Hood* by focusing on the impact of fear on one's psyche. He used narration and images to convey his entire story. The student also made a very interesting creative choice by coloring the entire comic in black and white with the exception of Red's cape, which was a fiery red. The cape leapt off the page since it was the only color present; this evoked images of distress, reinforcing the entire point of the comic. Another group of students chose to re-tell *Harry Potter* by focusing on an earlier generation of Hogwarts students, including Harry's parents and Severus Snape. This group focused on a single event in the lives of the Hogwarts students. This group was also fortunate to have an art student drawing their comic electronically. She created the comic entirely in black and white, and drew impressive likenesses of the characters. This particular comic, however, focused more on narrative than images. Another pair of students created a story about the creation of a Jamaican bobsledding team. This comic managed to strike a pretty nice balance between the verbal and the visual, and there were a couple particularly unique aspects to this comic. Parts of the narrative were written in Jamaican dialect and the students included Standard English translations at the bottom of the pages, which was a nice touch. The student who drew the comic used colored pencil for all of the text and images, making this one of the only full-color comics that was created in the class. Once again, multiple types of narrative and designs worked for this project.

For others interested in creating a similar assignment, I know the time and resources may not be available, but it would be relatively fast and inexpensive to professionally print personal comics for each student instead of doing a complete anthology for everyone. As evidenced by my experience, a large anthology can also be done for a fairly low price, making this project possible for any department with minimal funding. Anyone with access to color printers and scanners should be able to successfully print this project since it could easily be done on standard-sized printers using 11x17 inch paper, making the project even more accessible to instructors with minimal resources. If time for the instructor is an issue, the printing and reproduction of the comics are also tasks that students can work on in the classroom if they have access to computers; this would be a great way to extend the project over the course of the semester and allow student input regarding the layout of the collection. Finally, if a tangible, printed project is not possible for any reason, one could set up a blog or other website for the students to showcase their work and see each other's work, which was very important to the students.

As previously mentioned, the main goal of this assignment was to get students to prove that they had fully comprehended what they learned in the classroom by putting it into practice. This project certainly met that goal because most students were able to effectively utilize the primary customs of comic creation: use of paneling, effective use of closure, utilization of abstract images, creation of narrative and effective interrelation of the verbal and visual. Naturally, each student or group of students did not successfully demonstrate *every* aspect of comic creation, but most students were able to utilize several aspects of comic creation effectively.

I would encourage any teacher to try non-traditional projects whenever possible. They definitely bring excitement into the classroom. As clichéd as it may sound, my students' enthusiasm as they shredded open their comic anthologies and the looks of pure joy on their faces made all of the extra work of production worth the time, effort and headaches. Additionally, creating an opportunity for collaboration and fostering a sense of community in the classroom is extremely beneficial to the students who may not have such an opportunity in other classes. Perhaps one of the greatest compliments I received was from a student comment on an evaluation. One of my students commented that she had taken the class in order to break out of her comfort zone and that this assignment had done exactly that; she noted that she was happy for the experience. Teaching a course on the graphic novel for the first time and assigning a creative project for the first time also forced me out of a comfort zone and I too am happy to have had the experience.

References

Blankenship, Elizabeth. "Re: Graphic Novel Question." Message to the author. 12 Oct. 2011. E-mail.

McCloud, Scott. *Understanding Comics: The Invisible Art*. New York: Harper, 1994. Print.

CHAPTER SEVEN

GENRE FICTION AND THE ACADEMY: AN INTERVIEW WITH AUTHOR MICHAEL ARNZEN BY CHUN LEE

Michael Arnzen is a celebrated author of horror, and he often behaves the part of a writer of gruesome literature. If you ask him to recall the most disturbing thing he's ever seen for "research" he might tell you about the time he went to a cattle ranch in order to see inside the stomach of a living cow as it digests its lunch. But take him away from authors' conventions, Stoker Awards ceremonies and zombie walks, bring up pedagogy or literary theory at a quiet place to get a beer, and he will talk your ear off (if he doesn't tear it off). For his dual roles as author of popular fiction and professor of English and Division Chair of Humanities at Seton Hill University, I felt Mike was uniquely qualified to talk about the connection between the Ivory Tower and the world of genre fiction. In the following interview Mike describes a complicated, sometimes conflicted, but ultimately necessary relationship between the two worlds.

Does the Ivory Tower hate genre?

Hate is too strong of a word. What I think many academics in the proverbial Ivory Tower assume is that genre literature -- mass marketed entertainment -- is so preoccupied with getting an emotional response that it has nothing to offer the intellect. Thus it's more about the body than the mind, and this is why many call it a "guilty pleasure." Titles like *Fifty Shades of Grey* do nothing to dispel this myth. The sad truth of the matter is that the people who feel this way often are the people who never have read much of it, and they are basing these ideas on early experiences or even childhood memories of the various genres, and never bothering to give them a serious look.

Do you see movement from the Ivory Tower in regards to genre?

I do. It's a generational thing. I dare say all academics working today -- at least in America -- grew up with television and a world where big book chains categorized titles by genre. And every genre has appropriated the literary movements of the past in historical fiction and cinematic adaptations of classics. Perhaps it's a cliché, but the boundaries are blurring. Moreover, English graduate students are writing about everything from video games to films to internet websites. The notion of text is all-inclusive, and any scholar who thinks only canonical literature should "count" in English studies is really working on the margins anymore. Ironically, it may be popular culture that ultimately sustains the myth of the academic as ivory tower misfit.

What is your best defense of genre?

In terms of literary merit, one need only read widely in a genre -- including those titles in its history -- to see that things are not so black and white, and that not only is yesterday's genre fiction today's "literary" fiction, but also that there is a spectrum of literary qualities that stretches within every genre. We have "literary horror" as much as "splatter fiction" or "teenage vampire killer fantasy" for instance.

Are there works you would consider not worth reading or studying?

No. One shouldn't be selective, in my opinion. Our place in the world, our worldview, our subject position -- these things are always already limiting the scope of our awareness. Scholars seek to not only understand the world in their purview, but to also expand its horizons. You have to be indiscriminate about studying and reading the world around you to grow. I actually wish genre readers understood this as much as academics would. Genre boundaries should not be walls obscuring one's horizon line.

Why is there resentment from fandom of the ivory tower?

Well, this is a tricky question, because I don't think fans think much about the ivory tower at all. But there is a lot of anti-intellectualism in mass culture, and when we see it, it is often based in defensive ignorance. Professors are often mocked for being "nutty" or "absent-minded" or "pompous blowhards" and these stereotypes circulate often in popular fiction, too. I also think that fans may have some resentment of academia because they get no support from it -- pop books aren't archived in college libraries, reviewed in literary criticism journals or kept in print to serve courses in literature and things like that. We lose too many stories this way. But there is a degree to which fans ARE academics, in every way besides having the diploma to match it. Fan culture takes its genres very seriously -- serious fans don't just fetishize their texts by, say, buying related t-shirts and memorabilia -- they also often study them, write about them for the web or in fanzines or even in legitimately published books, and they also host discourse communities online. Heck, many "fan conventions" are only a shade away from being identical to a "literary conference." Academics are just like fans of whatever their specialization area might be. I wish these lines were crossed more often.

What is your favorite pop culture work to teach?

The Exorcist. The movie version. I've taught it in different contexts, from 'horror writing' courses to film courses to our survey course in 'literary criticism.' It was actually most effective in the senior literary criticism course, because it not only invited students to apply the theories they were learning to a fairly easy-to-understand film, but it also resists interpretation because it is so aggressively going after the emotions. I think the film is not just emotionally disturbing, but also intellectually challenging. I've written more articles about it than I think any other text, too.

Do you ever get resistance from students when you teach genre?

Sometimes. But they learn to play along because I invite all attitudes to have their say, so long as they are interested in sharing their reasoning, not just superficial judgments.

Are we too defensive of genre?

I'm not sure it's a battleground. But yes, sometimes some authors who cry out for legitimacy of genre "protest too much." I think the genre's best defense is good writing, and celebrating/sharing good writing by others.

Does genre limit the writer? Should writers leave marketing labels to agents and editors?

Writers should only think about 'labels' after their 'products' are complete and ready to package. To me, genre is more of a worldview -- a lens --

than a constricting structure that must contain a set list of ingredients. Readers who share my worldview will likely read the same genres as I do, so I like having the indexical term "horror" on the spines of my books, and in the search engines online. It is a community flag as much as anything else. And like any flag, it might stake a claim of a certain territory, but writers should be burrowing underground in their exploration, and maybe even moving the boundary line all over the map. It shouldn't matter to the author when they are composing. Genres matter more to the discourse community that talks about and reads these titles.

When we teach genre should we do something like consider the formula of the genre, the history of it or the marketability of it?

Genre should be operationally defined, and constantly open to redefinition and renewal. But teachers can't be so slippery. I always say that teaching brings the formalist out of us. We like to teach the patterns, the structures, the conventions, the trends. While I wouldn't use the term "formula" I would say that the traits that all genre texts share should be the starting place for the teacher. Then you use the history of the genre to test these patterns -- to see how different authors adopt or challenge the definition. And marketability? Well, maybe for creative writers who want to be freelancers. There's nothing wrong with scoping out the industry and analyzing how it impinges on or influences our literary understanding. Just as one could study syllabi or reviews or libraries, one can study the publishing industry and the business side of all this...because we're talking about forms of power, now, not just literary elements. But these are more rarefied topics that might be more teachable in upper division or graduate school courses. "Market" for early learners could mean more of an introductory look at the different ways that books appear in pop culture.

What did you learn about genre from editing Many Genres, One Craft*?*

That every genre writer is an educator on some level. Whether it's in the lessons of their narratives or the lessons they want to overtly teach others. Writers are smart, really smart, across the board, even if they write something you would presume is simple or easy-to-digest. I also learned about the diversity of genre literature, the wide array of issues and problems and beliefs that come up when you hear writers talk about their texts. Yet all popular fiction writers share a lot of the same struggles, and the only community they truly have is in each other. There's real community in *Many Genres, One Craft*.

You talk about labels in "Genre Unleashed." Do you wish there were no labels and the audience could just find your work without it having a label?

No, that would be like birthing people without giving them any names. I like to be able to look people up in the phone book. But I would never say, "Oh, you're a Quincy. I know Quincies... Quincies are idiots. Guess I'll be seeing you then." Labels can too quickly become categories and lead to reductive thinking. It helps the industry to index fiction and market it and shelve it and review it and study it and teach it. But it doesn't help us in reading it very much, and every good work of genre fiction both meets the label and transcends it. There's always a tension between a thing and its name, isn't there? I'm Mike Arnzen -- just like all the other Arnzen's -- but I'm the only Mike like me. When I go to the genre fiction shelves in the bookstore, I am looking for something that will both give me what I'm looking for (i.e., the genre conventions) and give me a total surprise (i.e., creative inventions). Every book has this dynamic struggle between the same and the new, but it really is most evident in genre fiction, because the expectations are right there on the table.

***

Michael Arnzen was a special guest at the Louisiana Conference on Literature, Language and Culture in 2012, where a different version of this interview was performed live. He is the co-editor of *Many Genres, One Craft: Lessons in Writing Popular Fiction* (Headline Books, 2011) Currently he serves as Division Chair of Humanities at Seton Hill University, where he has taught in their MFA in Writing Popular Fiction graduate program since 1999. Visit him online at http://gorelets.com.

CHAPTER EIGHT

MILTON TUBE: THEORETICAL IMPLICATIONS AND PRACTICAL USES OF *PARADISE LOST* WEB ADAPTATIONS

MICHAEL BERNTSEN

Exploiting the freedoms of online virtual communities and the technological advantages of the World Wide Web, amateur directors, graphic designers and comic strip writers have produced numerous adaptations of John Milton's works. YouTube, for instance, features numerous parodies, readings, and presentations of *Paradise Lost* for entertainment as well as educational purposes. The most popular ones include Patrick McCarthy's 3-D short animation film, "*Paradise Lost*: Book I," and Jon Roz's "War in Heaven and the Fall of Mankind" slideshow, each receiving thousands of views and containing obvious didactic implications. Another popular web adaptation is John Troutman's series of online *Paradise Lost* comic strips, published on the *Lit Brick* website. Despite Troutman parodying every aspect of Milton's epic, it remains a comprehensive adaptation that preserves the original narrative, characters and conflicts. McCarthy, Roz and Troutman take advantage of the web as a popular medium to present modern interpretations and presentations of this classic work. Their creative efforts not only participate in the act of adaptation, which includes theoretical issues that concern fidelity, authorship and genre tropes, but also demonstrate how *Paradise Lost* is a timeless work that contains relevant themes for modern audiences. Fluctuating between impulses to parody and retell *Paradise Lost*'s complex narrative, between entertainment and didacticism, between preserving Milton's voice and promoting their own, these contemporary adapters offer answers as to why such a bewildering and dense text continues to be taught, read and enjoyed.

In the contemporary virtual theater of YouTube, artists who use *Paradise Lost* as their subject usually do so for didactic purposes, in an attempt to aid students struggling with reading comprehension. Patrick McCarthy's short, "*Paradise Lost*: Book I," for example, is no exception, offering a computer animated short that depicts the epic's first lines within a running time of 7 minutes, 42 seconds. Featuring McCarthy reading passages combined with tableaux that contain passages from other sections, his short directly parallels much of the action after the Fall. McCarthy begins his film by showing a montage of Adam and Eve's plot points during the entire epic. As he reads lines from the poem, he depicts Adam and Eve enjoying paradise, Eve's seduction by the serpent, and their expulsion. This focus reinforces that the beginning of Book I provides evidence that establishes Adam and Eve as the epic's tragic figures, not Satan. Although McCarthy's dramatic nexus is Satan's transformation from angel to demon and his monologues in Hell, granting Adam and Eve screen time reveals McCarthy's interpretation of the epic. Like many other adapters and critics, McCarthy is fascinated with Satan as a character, yet his need to begin with a brief relation of Adam and Eve's narrative is a testament to Milton's craft for characterization. Satan is certainly an interesting character, if not the most interesting one, yet McCarthy identifies Adam and Eve as the epic's emotional center. Their loss is the greatest.

McCarthy's choice to include Adam and Eve reveals how adapters undertake dual roles in the creative process by playing both artist and critic. One continuing debate among Milton scholars is the determination of the epic's tragic hero. William R. Herman, in "Heroism and *Paradise Lost*," John M. Steadman, in "The Idea of Satan as the Hero of *Paradise Lost*," Stella P. Revard, in "Eve and the Doctrine of Responsibility in *Paradise Lost*" and numerous other critics past and present partake in the controversy of deciding who is Milton's intended hero. At the immediate stage of the adaptation process, McCarthy joins in this academic discussion by providing his interpretation on screen, reinforcing how adaptation is an art form of revision, homage and interpretation. What an audience hears through an adaptation is a blending of voices that address critical and personal concerns of the adapter and the prime author.

Determining how the poetic voice of *Paradise Lost* should sound remains a challenge for any adapter who wishes to reinterpret, revise, retell or reinvent the epic. Even John Dryden took issue with Milton's language, yet the poetic style entices many readers and scholars to the work in the first place. Although McCarthy intends this project to help anyone suffering from comprehension issues, he preserves Milton's original language throughout most of the film. He hopes that the addition of visual cues will decrease the density of Milton's text. The only moments in which he veers a bit from the actual text is the dialogue between Beelzebub and Satan, changing "If thou beest he; But O how fall'n! how chang'd / From him, who in the happy Realms of Light" (ll. 84-85) to "Oh how I have fallen so far, once bright in the heavenly realms," for example (McCarthy). Although he updates the diction, McCarthy preserves the poetic rhythm because his film relies on the original text and must maintain this tone and sound to remain consistent. McCarthy offers a reason why he paraphrases this exchange, which is "to help readers be aware of what is physically going on during that scene, so that when they read back, they can pick up all the small details of Milton and be aware of what the exchange between Satan and Beelzebub is all about" (McCarthy). In his justification and actions, McCarthy echoes the decisions all adapters must make in determining what to preserve, what to revise and what to reconcile.

Since McCarthy is concerned with readers comprehending Milton and internet audiences enjoying his adaptation in conjunction with the original text, his short film highlights the collaborative nature interwoven within all adaptations. McCarthy's exercise in adaptation is not a strict authoritative assumption of Milton. By including Milton's exact lines, McCarthy simultaneously calls attention to the epic as well as his own adaptation. What he accomplishes is a balance of authorial wills, respecting Milton's vivid language and adjusting it when he needs to sustain his own dramatic force. This adaptation, designed for pedagogical purposes and not meant to be the endpoint of understanding Milton, can help dismantle any confusion about the first book, introduce people to Milton's poetic language, and, hopefully, inspire unfamiliar audiences to read the epic.

Jon Roz also presents his adaptation, "War in Heaven and the Fall of Mankind," as a slightly educational tool, emphasizing that it is "loosely based" on *Paradise Lost* (Roz). His short runs 3 minutes and 8 seconds and includes more text on the screen than McCarthy's video, which creates a clear attachment to the literary origin and certainly lends it a didactic aspect. Arranging most of Gustave Doré's illustrations into a storyboard depiction of Books V and VI and mixing them with direct quotes from the epic as well as his own abbreviated versions of passages, Roz's adaptation presents a more profound issue concerning fidelity than McCarthy's short does. Roz disregards Milton's narrative progression and realigns Doré's illustrations with passages that do not match the original order. Roz's choice to begin with the war instead of the Fall originates from a previous project, in which he depicted only the war in heaven because it is his favorite part in *Paradise Lost* and in Doré's collection. After completing the first video, however, he wanted to tell the entire story, albeit condensed, because his love of Milton prompted him to be more faithful to the text. He could have inserted scenes but decided to tell the epic from the real beginning to its end because he felt flashbacking would be too confusing, given how he depends upon still images and brief passages of text rather than moving characters and dialogue to generate story cohesion (Roz). The fact that Roz is not faithful to Doré's adaptation further distorts fidelity and demonstrates the impossibility of strict adherence to any prime source. The moment a reader opens a book, fidelity is tested, if not lost. Once that reader turns into an adapter, the major concern is not loyalty to the original source, but the successful expression of one's own vision and interpretation. That aspect, however, may not be possible since what an artist intends and what an artist produces are sometimes at odds.

Roz's focus on Milton's text coupled with the images also demonstrates the compromise and inherent sacrifice visual adapters have to undertake, especially for pedagogically grounded enterprises. In a visual adaptation, the options of sight and sound offer unique representations of the text in order to solidify textual ambiguities, yet even Roz excludes much of Milton's complete passages concerning the war in heaven because visual media naturally move away from strict textual descriptions. Roz's video aims to create a more sensory-based entity from which the original text can operate, fulfilling its artistic intentions, and from which more visual-based learners can gain an understanding of the literary work. Roz fears, however, given the free access to YouTube, that students would substitute watching his or other people's versions of it for reading the epic (Roz). His concern addresses how adaptations can disrupt the collaborative process from which they emerge. Watching the movie version instead of reading the book is not an abuse of the process, especially if each work is an independent piece, yet the relationship between each work is severed. This issue is not one of fidelity, but of connection. It is the anxiety of obscurity. Adaptations are not meant to replace works that need forgetting; rather they are meant to reintroduce previous works and inspire future ones.

These two short films demonstrate the intertextual collaboration inherent in the adaptation process. Despite the vast technological resources available through this visual medium, McCarthy and Roz can never fully escape Milton's original text since each one uses text on screen, preserving the literary aspects to the epic. Since the prime source is a text, the visual and audio based adaptations still echo the origin point, serving as visual learning devices for those readers who have difficulty with strict textual comprehension. Thomas Leitch argues that the literary collaboration involved in the process of borrowing and transferring one text to other media is the consequence of, and reason for, the existence of all adaptations (150). The collaborative nature of adaptation includes a profound attachment to the text and the author's original vision as the current reader, or future adapter, sees it. Stressing the intertextuality that exists in all literature, Leitch claims that "[a]lthough it is certainly true that adaptations are intertexts, it is equally true that their precursors are intertexts because every text is an intertext that depends for its interpretation on shared assumptions about language, culture, narrative, and other presentational conventions" (166). An expanding chain of reader, text and author relationships emerges as the act of adaptation begins. What occurs during the adaptation surpasses the traditional reader-author-text triangle. All three forces participate in the creation of a new text that convolutes the normal transactions within the triangle since the reader becomes an author and a text begets another text that will occupy a point in a separate triangle but contains shadows of the first text.

Harold Bloom suggests that this collaboration can be overwhelming to artists rather than a positive engagement between prime author and text. According to Bloom, artists are in a constant negotiation of suffering, inspiration and independence whether adapting or creating prime sources. Milton, in his estimation, "had to struggle, subtly and crucially, with a major precursor in Spenser, and this struggle both formed and malformed [him]." (Bloom 11). Successful artists, like successful adapters, simultaneously suffer and embrace the history of influence before them. Following Bloom's postulations, the act of adaptation might not necessarily be a collaboration but a conflict of interests and authority. It is a documented account of the poetic influence, one in which the adapter's authority and originality overlap or subdue or are conquered by the prime author.

Roz and McCarthy must depend on text, or the visual representation of words, even though their adaptations are not bound by pages because the literariness connects their works to Milton's epic. Of course, this dependence might not reveal "the anxiety of influence," as Bloom suggests so much as it articulates an active engagement between adapter and author. The fact that Roz uses Gustave Doré's illustrations based on *Paradise Lost* in his own version demonstrates that this process is a dynamic recognition of past interpretations as well as a presentation of an adapter's unique vision. Doré does not inhibit Roz's voice; he, in fact, adds more visual expression to what Roz wants to capture artistically. Rather than the pure anxiety that Bloom obsesses over, adaptation functions as a blending of inspiration and interpretation.

Linda Hutcheon addresses the issue of fidelity by arguing that adaptations are independent "aesthetic objects" that may be haunted by the original entity, but not damned by it. Hutcheon asserts that adaptation is not replication and obliterates any sort of fidelity criticism by stating that "proximity or fidelity to the adapted text should not be the criterion of judgment or the focus of analysis" (6). The relationship to the root source will guide audience interest and critical reception; however, the anxiety over fidelity should not plague the way in which scholars approach the study of adaptation. Through the process of adaptation, during its creation and after its completion, the original source material embodies a different life, a different artistic space. Although changes are inevitable, the spirit of creation unifies any adaptation with its prime source. This outcome suggests that the power of adaptation grants active, communal participation, offering the possibility for artistic immortality to both works and creators.

John Troutman with his virtual comic strip parody of *Paradise Lost* attests to how adapters work in close proximity with the source text as well as enjoy the artistic collaboration that the process of adaptation offers. Unlike McCarthy and Roz, Troutman employs a comedic tone to the tale, yet he still cannot escape Milton's authorial ghost, or the academic debates surrounding the epic, since he provides brief commentaries under each strip that discuss where and why he departs from the original text. Even as he pokes fun at Milton's writing and storyline, Troutman simultaneously participates in critical debates concerning character and poetic interpretation. By blending illustration with text for his version of *Paradise Lost*, Troutman reveals how the act of adaptation parallels the art of translation as he transforms a canonical work into a contemporary and technologically-based piece of art.

After parodying works from Chaucer, Marlowe, and Shakespeare on his LitBrick.com homepage, Troutman decided to approach Milton's epic simply because he is following the chronology of *Norton's Anthology of British Literature* (Troutman). Having never read *Paradise Lost* before and ignoring the Norton's companion summaries and criticisms, Troutman does not have the same relationship with it, as McCarthy and Roz have, prior to his adaptation. In fact, Troutman prefers this freshness of approach in order for "the comics to reflect [his] own, often completely wrong, interpretations of the text. While critical analysis would probably make [his] comics and commentary more accurate, [he] finds that [his] own ill-informed conclusions are considerably funnier." Unknowingly, Troutman's methodology for adaptation resonates with other *Paradise Lost* adapters and adaptation theorists before him. Troutman aims to be an independent artist throughout the process of adaptation. His authorial vision and interpretation guide his work, creating a new piece of art that is separate from Milton's work. Of course, even though he neglects criticism, the anxiety and the long tradition of influence reside within his work nonetheless.

Despite his lack of familiarity with the text, Troutman's adaptation, parody as it may be, is quite faithful to Milton's original storyline and characterization, which indicates that parody further complicates issues of fidelity. When parodies function as discernible adaptations instead of works that merely contain references, fidelity is important to the overall comedic success. Troutman's compulsion to follow the narrative reveals how the epic's conflicts and characters continue to captivate readers of the text and corroborates what Samuel Johnson means when he utters his high praise of Milton's narrative being so tightly constructed. Acting almost as a warning for those who consider dismantling or rearranging Milton's epic, Johnson comments that "[t]here is perhaps no poem, of the same length, from which so little can be taken without apparent mutilation" (304). Although Johnson hints to editorial butchery rather than adaptation projects, any artist attempting to adapt *Paradise Lost* encounters this dilemma and hears Johnson's haunting words. To adapt this work is to reduce it; however, such an act is not one of infidelity or insult. It is a reminder of adaptation's power, merit and relevance as an art form.

As an unintentional embodiment of Johnson's critique, Troutman preserves many of the major narrative elements because Milton has crafted a well-structured plot, which includes characters and situations that are compelling for any era. Unlike some of his other strips in which he satirizes or parodies particular scenes, Troutman does not begin this project until after he spends time outlining the plot to all twelve books then considers which major points should remain. This paradoxical fidelity to the text, his mixture of preserving Milton's major narrative points with Troutman's need to decipher what those points are and condense what is not helpful for his adaptation, indicates that the basic conflicts are universal to any century and that the characters remain relevant reflections of our humanity and personal struggles. Even when an adapter injects humor into these elements, their core relates to audiences and the familiarity with these tragic figures inflates the comic presentation.

The way in which Troutman depicts Eve's temptation, for example, still follows Milton's argument for why Eve would eat the apple, but accomplishes the characterization and dialogue between Satan as the serpent and Eve in four frames:

[1]

Playing upon Milton's Eve, who has proposed to Adam that they separate in order to hasten their upkeep of Eden, Troutman has his more modern Eve vividly frustrated with her chores. Here, the serpent talks in a friendly tone and presents a fallacious argument grounded in common sense. Sensing her lack of work ethic, Satan purposes that eating the apple will make her life easier. Troutman inserts his commentary on Eve's passivity in Milton's epic by adding the line, "and I'm s'posed to do whatever Adam says," which harks back to one of the main dilemmas of modern adapters of *Paradise Lost*. Unlike Collier who hopes to make his Eve more assertive and have a higher narrative presence, Troutman satirizes Eve's submissive station. His focus on the apple making Eve's brain work better and Eve's observation that the bite has "brought her science," presents a strong will despite her childish nature since she desires to explore the world around her. Troutman's Eve, despite her improper grammar and adolescent demeanor, is as logically driven as Milton's Eve. Her choice to eat the apple is based on good intentions even though they are selfish. Concluding with Eve's realization that she's naked is comical, yet the realization still strikes a tragic chord. Even presented comically, the concept of seeing nudity as shameful reinforces the universal yearning that innocence should not have to disappear with experience and knowledge.

Troutman's adaptation, too, is the most modern in its visual presentation and language, especially in comparison with the YouTube videos. Bringing a modern look to the epic, he has Uriel wear blue jeans and no shirt as if he were a Calvin Klein model, Sin sport a gothic, nerdy look with green tentacles for legs, and Eve looks as blonde and ditzy as possible. These modern aspects are immediate in the first strip:

---

[1] Troutman, John. "*Paradise Lost* 16." LitBrick.com 19 September 2011 Web. 16 November 2011. Reprinted with permission from Troutman.

2

Troutman's choice to present the demons as mostly regular looking people with modern styles suits his comedic purpose and enhances familiarity with modern audiences. Although Troutman admits that he "didn't feel like drawing the traditional devil," these depictions are important for the metaphoric relation of Milton's themes. Having Satan look more human blurs the differences between him and humanity, heightening the danger of his character, which Milton certainly makes clear in the original epic. The conversational diction also adds helpful rhythm and characterization. Like Eve's improper grammar, the demons' level of diction fits their intelligence and provides quick humorous setups for Troutman. Their concerns, in this case, for a waterslide and the idiocy of a golden palace in hell reflect Troutman's decision to use popular culture as a forum for humor and to include small tidbits of information for educational purposes.

Despite shaping his adaptation of *Paradise Lost* as a parody, Troutman clearly intends it for an educational purpose as well as an entertaining one. He has clear missions with his adaptation, which is to modernize them by presenting web adaptations and to extend serious literary works into the realm of comedy. His adaptations involve critical interpretations of the works as well as Troutman's aesthetic choice to use comedy to reintroduce classical texts to contemporary audiences. Although he enjoys bringing humor into his adaptations, he explains that "[a]n added side benefit to poking fun at these works is that I can help draw attention to more obscure stories that most readers may never discover themselves." Even though he figures most people would be familiar with *Paradise Lost*, he provides a commentary underneath each strip that summarizes plot points and what he has altered in order to explain to people unfamiliar with the work what he is exactly poking fun at. These sections "frequently give [him] an excuse

---

[2] Troutman, John. "*Paradise Lost* 1." LitBrick.com 29 August 2011. Web. 9 September 2011. Reprinted with permission from Troutman.

to crack additional jokes at the work's expense," which shows how adaptations can serve as an invitation to the original source material. Troutman's use of comedy is an integral component to demonstrate that the characters, plots, and themes of classic works are relevant, relatable, and translatable to modern sensibilities. In order for the audiences to be fully in on the joke, they need knowledge of those works; thus, Troutman promotes serious literary interest as he parodies and satirizes works on the highest level of the literary canon.

One particular point at which Troutman, McCarthy and Roz encounter the limitations of their craft is the need for traditional representations of shapes and objects. Since each one is not working with abstract forms, they rely on conventional and recognizable imagery. Milton's black flames of hell is one concrete image from which all three illustrators must depart because depicting it would offset their designs and require too much inference from the audience. Each adapter uses the common colors of flames, yellow, red, and orange, in order to produce identifiable associations of hell and fire so the location is immediately clear for the audience. Preserving Milton's description of Hell's black flames would be detrimental to Troutman's more cartoonish medium because they would be too odd for comedic effect, not to mention blend in with the demon's hair. In the process of adaptation, clarity and convention sometimes guide artists away from authorial fidelity because certain ideas work better in specific media.

McCarthy, Roz and Troutman represent only three of the numerous visual artists who have adapted scenes from *Paradise Lost*, demonstrating the plethora of images that beg for visual reorientation from another medium. Discussing the debate concerning whether Henry Aldrich or John Baptista de Medina were the first to design illustrations for the 1688 edition, John T. Shawcross relates how their particular styles and their era's aesthetics guided their interpretations of Milton's images (43). Visual adaptations are then determined not only by personal interpretations but cultural taste and expectations. Shawcross' investigation also shows how immediate artists foster the relationship between *Paradise Lost* and the visual arts, which reveals how naturally Milton's imagery translates to visual modes. These investigations, conducted by many other critics as well, suggest that a good work prompts a continual adaptive collaboration. This process produces innumerable presentations of a single line or image or idea, making visual adaptations more than helpful teaching aids as they test the boundaries of originality, creativity and adaptation.

One current that certainly flows through the adaptations of McCarthy, Roz and Troutman is the didactic spirit. While each artist is having fun with these projects, they recognize that the main reason most people will view their adaptations is due to comprehension issues because of the general resistance to *Paradise Lost* even from professors. To teach such a monumental literary endeavor, of course, presents abundant challenges. Jeffrey Shoulson, for instance, describes bringing the epic into to the classroom as an "intellectually and pedagogically exhausting experience" (189). Wendy Furman Adams states that "twenty-first century college students often are thoroughly daunted by the manifold complexities of Milton's conceptual universe—political, theological, ethical, and linguistic" (62). Given the common linguistic infidelity, all adaptations of *Paradise Lost* provide a substantial arsenal in counteracting these obstacles.

Nevertheless, bringing in adaptations can pose another set of obstacles to the learning process because this tactic carries insinuations that the original work is irrelevant or unpleasant. If works like *Paradise Lost* require adaptations to enhance their impact with students, they might wonder why instructors are not basing their courses around the adaptations. Dennis Cutchins, Laurence Raw and James M. Welsh, who encourage the use of adaptations in classrooms, urge teachers to discuss adaptation theory even briefly in order to discuss the relationship between a source text and its progeny. Since artists engage in critical analysis during the adaptation process, introducing adaptation theory can illuminate the presentation of such works so it does not seem like adaptations function as supplemental material. Cutchins, Raw and Welsh argue that the most effective way of using adaptations in the classroom is "to teach adaptation by having students read a great deal of literature and view a great many films. But this needs to be *guided* viewing and reading in which students are taught to *think* and to *see* in the manner of adapters" (xv). Despite their ideal situation not accounting for time allotted to teaching *Paradise Lost* in the usual college course, their call to foster a perspective that adaptations function as critical and artistic expressions is important for students to understand why adaptations are independent works that function as pedagogical tools to benefit its source text and as prime texts begging for its own critical scrutiny.

Adaptations do not exist so teachers can take a day off of reading or preparing. They also do not exist so critics can simply exercise their compare and contrast skills. The creative efforts of McCarthy, Roz and Troutman reveal that adaptation is a complicated dialogue between the source text, the original author, the new text and the adapter. Adaptations, like any artwork, reveals contemporary tastes and historical evolutions of aesthetics. Beyond theoretical concerns, these adaptations further offer a fresh dialogue with the primary material and can serve important practical and pedagogical purposes by aiding in the appreciation and comprehension of the original work, especially since most students need a lot of help appreciating and comprehending *Paradise Lost*. Milton's presence on the web presents an important nexus that reflects how adaptation preserves classical works and creates new, independent pieces of art.

In fact, every adaptation of *Paradise Lost* is a testimony to the epic's universal themes and wide-ranging appeal. These adaptations are also important, creative moments that testify to the impact, inspiration and power of literature. Although we are far removed from the Greeks who passed their stories through oral traditions, we continue in the epic ritual of storytelling through the practice of adaptation. *Paradise Lost* is a story worth retelling and revising because it contains many elements that help make stories so captivating: an origin myth, great sacrifices, tragic loss, moral lessons and a hopeful interpretation of God and the future. Balachandra Rajan describes the epic as "a poem of the infinite theater, not in the sense that it is a drama grown monstrous, but in the sense that the massed resources of time and space are made to move inward on the moment of crisis, intensifying that crisis and to some degree measuring it be its endless involvement in a web of meaning and consequence" (44). In justifying God's ways to man, Milton composes a work that has the ultimate human drama at its center and justifies why two people would abandon God. The consequence of Adam and Eve's defiance is the knowledge of all that is mortal and the full culmination of all that is human. *Paradise Lost* offers a reason why a man would sacrifice his god for his wife. It offers an explanation of our world and a reminder that humans endure the hardships and celebrate the beauty of a world that seems to revolve without divine intervention. Milton's masterpiece will endure so long as people stretch their dreams beyond the scope of our transient realm.

References

Adams, Wendy Furman. "'That Glorious Form': Teaching the Nativity Ode and *Samson Agonistes* through Visual Images." Ed. Peter C. Herman. *Approaches to Teaching Milton's Shorter Poetry and Prose*. New York: The Modern Language Association of America, 2007. Print.

Bloom, Harold. *The Anxiety of Influence*. New York: Oxford UP, 1973. Print.

Cutchins, Dennis, Laurence Raw, and James M. Welsh. *The Pedagogy of Adaptation*. United Kingdom: Scarecrow Press, Inc. 2010. Print.

Hutcheon, Linda. *A Theory of Adaptation*. New York: Routledge, 2006. Print.

Johnson, Samuel. *The Lives of Poets. The Works of Samuel Johnson Volume 10*. Edinburgh: Bell & Bradfute, 1806. Print. 304.

Leitch, Thomas. "Twelve Fallacies in Contemporary Adaptation Theory." *Criticism* 45.2 (2003): 149-171. Print.

McCarthy, Patrick. "*Paradise Lost*: Book I." YouTube. 7 November 2008. Web. 4 April 2010.

Milton, John. *Paradise Lost*. Ed. Judith A. Jance. New York: Avon Books, 2002.

Rajan, Balachandra. "*Paradise Lost*: The Hill of History." *Huntington Library Quarterly* 31.1 (1967): 42-63. Print.

Roz, Jon. E-mail interview. 7 October 2010.

—. "War in Heaven and the Fall of Mankind." YouTube. 18 November 2009. Web. 10 September 2010.

Shawcross, John T. "The First Illustrations for *Paradise Lost*." *Milton Quarterly* 9.2 (1975): 43-46. Print.

Shoulson, Jeffrey. "'Passion Spent': Teaching *Paradise Regained* and *Samson Agonistes* as Twin Texts." Ed. Peter C. Herman. *Approaches to Teaching Milton's Shorter Poetry and Prose*. New York: The Modern Language Association of America, 2007. Print.

Troutman, John. E-mail interview. 29 September 2011.

—. "*Paradise Lost* 1 & 16." LitBrick.com. 19 September 2011. Web. 16 November 2011. Reprinted with permission from Troutman.

PART IV:

BEARS, SHARKS AND VAMPIRES IN THE POSTMODERN WORLD

CHAPTER NINE

THE FRENCHMAN AT THE KITCHEN TABLE: THE INFLUENCE OF JEAN BAUDRILLARD, ROLAND BARTHES AND JACQUES DERRIDA ON BACHELDER'S MODERN AMERICAN LITERARY FAMILY

KRISTEN GIPSON

The modern American sitcom family sits down to a round kitchen table where everyone from the head of the household to the family dog shares their opinions about politics, religion and world events over a well-balanced meal. Everything happens following a linear, chronological plot. Irony, much less satire, does not exist. Language is as upfront and straightforward as the family itself. The expectation of the modern American family is always set and met with expediency and cheerfulness: this is the *mythologie,* the simulation, the sitcom family through which we are all inevitably disappointed, because it is not attainable. It is more real than real. Upon closer inspection, the joyful binaries of the sitcom family—man/woman, husband/wife, fidelity/infidelity—will always deconstruct because they are simulations: why isn't Mike Seaver helping Tracy Gold with her eating disorder? Why does Mrs. Cleaver have to vacuum in *heels*? Questions that force their way past the simulation also question the expectations of the American family, and whether these expectations can, in fact, guarantee happiness. Modern American literature has long engaged with the fact that there is something inherently wrong with the expectations of the American family. In particular, those problems seem to revolve around that the American family is not, as advertised, stable, because it is a simulation, an illusion. Any family can achieve sitcom family status, but only with a script in hand, with emotions in check and with low expectations. The theories of simulation, simulacra and deconstruction as discussed by the French theorists and philosophers Jean Baudrillard, Roland Barthes and Jacques Derrida particularly inform the culture of the modern American family. For examples of the modern American literary family, we will look to three stylistically distinct authors; particularly, on Kathryn Davis's *Hell,* Chris Bachelder's *Bear v. Shark,* and Julianna Baggott's *The Miss America Family*. Each of these novelists uses obstruction of language, deconstruction of binaries and simulation in order to expose the simulated, sitcom family for what it is: a hollow representation.

Perhaps the best way to prelude a discussion of simulation, *mythologie* and *difference* is to respond to Chris Bachelder's name-dropping invitation in *Bear v. Shark*. Though many authors of note, philosophers, influential writers and critics are mentioned, the mention of myth is perhaps the best entrée into a larger discussion, because it influences everything. In the chapter entitled, "Non Sequitur," Bachelder mentions that "Roland Barthes (author of *Mythologies*) says, In fact, nothing can be safe from myth" (Bachelder 120). What Bachelder does not mention is why. Barthes argues, in *Mythologies*, that all language is myth; in fact, myth itself is a type of speech (Barthes 109). Though there are conditions that must be met for something to become a myth, they are arbitrary conditions; thus, myth becomes a metaphor for deconstruction itself. If, as Barthes argues, he gives a woman a bouquet of roses in order to symbolize his passion for her that is not because the roses themselves are passion, but because they have been attributed (through mythology and through language) the symbology of passion. In the same way, a bear does not equate with anything but bearness and even the word "bear" gives an arbitrary, assigned mythology that, were a culture to change, would change the mythology of the word itself (and perhaps even the mammal itself). Therefore, modern American authors like Bachelder, Davis and Baggott will use the mythology of language and culture to show a break-down of not just language or meaning, but of society itself. As Jean Baudrillard says in "The Implosion of Meaning in the Media," "Myth exists, but one must guard against thinking that people believe in it: this is the trap of critical thinking that can only be exercised if it presupposes the naivete and stupidity of the masses" (81). In other words, we can see evidence of the deconstruction of language, culture and society, but often characters are ignorant to these happenings—it is the work of the reader to recognize the deconstruction as it happens and riddle out what meaning, if any, can be gleaned from the text.

In addition to manipulating language to provide an intrinsic discussion of mythology, modern American authors also frequently employ a battle of content vs. form in order to engage the audience into the discussion of simulation, which is itself highly concerned with how information is disseminated. A perfect example of this is Kathryn Davis's *Hell,* which is dizzying in its attempt to blend three narratives; to say that this blend is done seamlessly would be, perhaps, a stretch, but it is done and with apparent purpose. In fact, the novel itself would appear to revel in the seams, in the disharmonies that it produces. It would *not* be a stretch to assume that the main objective of *Hell* is to challenge the importance of content as compared to form: "Sooner or later, the house will get the best of you. It will defy your attempts at narrative because it's opposed to content; it only honors form" (Davis 23). The house is the house of the narrative, the dollhouse of the narrative, and the cottage of the past, but it is also the novel itself, and it is also none of these things, but media-at-large. At any rate, the message of the text seems to be that form is, if not superior, at least almighty, compared to content. However, while this is in keeping with Barthes—"The emptying out of the interiority to the benefit of its exterior signs, this exhaustion of content by the form, is the very principle of triumphant classical art" (Barthes 10)—it seems counterproductive and demeaning to the book itself. Though *Hell* is a classic lesson in deconstruction, it does not necessarily seem to have "no point," the way a book that exists solely within the realm of form does. Something which only honors form, not content would be the equivalent of a fluffy summer blockbuster, all explosions, empty banter and CGI. The content is there, but it is exhausted. In *Hell,* there is not only plenty of opportunity to burrow past the form in order to decipher the content, but this seems to be the *point* of the form itself. In summary, if the form exists primarily (or even secondarily) to highlight content, content is not exhausted but exalted.

In *Bear v. Shark,* form and content are dealt with in keeping with Baudrillard's concerns about information dissemination, a central theme of Bachelder's text. In the book, mass media circulates the Bear v. Shark dilemma to the point that it loses all meaning. The chatter of television is as omnipresent as the dialogue between characters. In the extra-textual culture (that is, American culture), media is given the appearance of being *either* more truthful, factual or authentic than interpersonal communication or less truthful, factual or authentic. In other words, because we acknowledge that the average American's life does not involve car chases, aliens or espionage, we acknowledge a Michael Bay movie as less authentic than interpersonal communication which we encounter outside of the movie theatre (the hitch in this theory is, of course, the flood of essays and reviews and dissertations, which may suggest metaphors in which aliens are, in fact, communists, so that a Michael Bay movie is suddenly transformed into *more* truthful than extra-cinematic communication).

On the other hand, generally, newspapers and television news programming are often seen as less biased and thus more truthful than interpersonal communication because they have the use of sound bytes, press conference footage, and so on to support their claims, which are delivered by well-coifed, eloquent and educated (and thus trustworthy) men and women. However, the problem in assuming that any type of information dissemination (i.e., media, television, science) is privileged over another type of information dissemination (i.e., rumors, gossip, folk tales, etc.) is faulty in that it assumes that the form—how the information is distributed—is more important the information itself. Or, as Baudrillard puts it, "It is useless to dream of a revelation through form, because the medium and the real are now in a single nebula whose truth is indecipherable" (Baudrillard 82-3). Thus, when Mr. Norman, Bachelder's confused protagonist, listens to the radio, of course he might be led to revelations not necessarily through form or content, but a combination of each:

> A woman is talking about a product. Mr. Norman cannot hear the words, but he recognizes the intonation of the sell. And maybe thïs is what it comes down to, not content but form—an Olympian's sleek body, a well-known actress's voice, a jingle, a striking use of color, a fade, a jump-cut, a swish pan, sound effects, a nice ass, innovative text-image placement. (Bachelder 171)

Succinctly, Bachelder undermines the argument which he has just presented. Though much of *Bear v. Shark* is an examination of the effectiveness of the form of satire, the most impactful thing satire can do is stress both the presence and absence of the thing which it is satirizing, in this case, media. Mr. Norman may come to the realization that form is more impactful than content, but in doing so, he also forgets to be impacted by the form itself; instead, he comes to a revelation of content itself. The lack of meaningful content is what the reader (and what Mr. Norman) notices, and is thus influenced by in this passage, and throughout *Bear v. Shark*.

Just as the theory of deconstruction suggests, the repeated use of certain words, the inflation, in a sense, of certain words, serves to devalue them, ultimately pointing to the ability of modern society to devalue language (and thus the meaning of language) itself. For Derrida, the use of certain words (or perhaps any words) is an invitation to deconstruct truth:

> Further [a logos] inaugurates the destruction, not the demolition but the de-sedimentation, the de-construction, of all the significations that have their source in that of the logos. Particularly the signification of *truth*. All

> the metaphysical determinations of truth, and even the one beyond metaphysical ontotheology that Heidegger reminds us of, are more or less immediately inseparable from the instance of the logos, or of a reason thought within the lineage of the logos, in whatever sense it is understood. (Derrida 10)

This devaluation is, in a way, a simulation. Language becomes a sign for something that does not exist; the language simulates a meaning that is not there. Or, "Rather than creating communication, [the social organization] *exhausts itself in the act of staging communication*. Rather than producing meaning, it exhausts itself in the staging of meaning" (Baudrillard, 80, original emphasis). In *Bear v. Shark,* the use of simulation in language deconstructs over the course of the novel, particularly when it concerns the fight, Bear v. Shark. On the most basic level, the word "bear" is meant to represent a mammalian creature, usually but not always indigenous to wooded areas. On the most basic level, the word "shark" symbolizes a large, carnivorous fish which has been known to occasionally attack mammals. On the most Derridean level, we must also assume that the word "bear" represents every word we relate to "bear"; thus, the word bear is also associated with the polar ice caps, global warming, Eskimos, forest fires, and so on. The word "shark" also assumes the associations of the movie *Jaws* and so on. Also following deconstructionist theory, there must be some sort of binary opposition that takes place between bear and shark. Here enters Bachelder's Japanese kōan [1] method of engaging the reader in the act of deconstruction. Normally, either a text or a societal dictum will indicate that one element in a binary opposition is privileged over the other: light/dark, good/evil, civilized/barbaric, for example. By implying that there is a competition between bear and shark, Bachelder suggests that there is one necessary privileged element of the binary, when in fact, there is no societal or natural basis for such a binary to exist. Thus the reader, like the characters of the novel, is constantly engaged in an impossible act of deconstruction. To further the dilemma, the words "bear" and "shark" themselves are not reliable, even on the most basic level, as they are stand-ins for something that both does and does not exist.

In *Bear v. Shark,* the "bear" and the "shark" are both natural creatures, in that the reader of the novel is no longer able to distinguish between the

[1] Japanese kōan: A practice of Zen Buddhism in which a story is told or a question is asked in which the meaning cannot be understood through rational thought but instead must be intuited. Or, "a paradoxical riddle presented to Zen students to help them break free of reason and, in essence, achieve enlightenment" (Bachelder 54).

real and the simulated which causes the simulated to equate to the real; complete simulations here references the fact that nothing like them exists in nature. The Normans, like their counterparts in this textual society, are not able to differentiate between the real and the simulated. When Mrs. Norman plays a virtual knitting game, she cannot see any element of ludicrousness in the act, because she does not see any difference (or difference has ceased to exist) between actual knitting with yarn and needles and virtual, or simulated, knitting. The act, though satirized to be ludicrous, is not unusual or absurd to the Normans, which means it is not absurd. Similarly, when the word "bear" is uttered in *Bear v. Shark*, it both signifies and does not signify the living creature which exists as a course of nature. On one hand, the Normans and their counterparts in society do seem to be engaging in a riddle in which "real" bears and sharks enter, as is evidenced by the author's poll of experts of biology and natural science. These people are discussing the likelihood of whether a bear, of nature, can fight or even win against a shark in nature. They are basing their argument, to some extent, within reality. They do not view this riddle as a simulated event, or an event in which bear-like creatures are facing shark-like creatures. Yet, this is exactly what is physically happening. Though the announcer at the Bear v. Shark event catalogues the "bear" competitor as "'a large mammal of the family Ursidae, found almost exclusively in the Northern Hemisphere, with a large head, a bulky body, short and powerful and clawed limbs, and coarse, thick fur, standing almost nine feet tall and weighing more than 1,500 pounds, Shakespeare's darling, well rested and looking for revenge…the *BEAR!*'" (Bachelder 248-9). In actuality, this bear is manufactured. It is *not* a mammal of the family Ursidae, but a robot made to look exactly like the creature is meant to represent. As in any aggressive sporting event, or, "As in the theatre, each physical type expresses to excess the part which has been assigned to the contestant" (Barthes 17). It is not found exclusively in the Northern Hemisphere, unless one takes this to mean that there are no Bear v. Shark competitions in the Southern Hemisphere. Moreover, this robot, though it may be indistinguishable in appearance from a "real" bear, does not employ any of the characteristics of a bear found in the wild—this bear has never scavenged for food, felt fear or suffered an injury, for example. It is more real than real, a kind of perfected reality. This creature, an utter simulation, is not the bear it proclaims itself to be; yet, *at the same time*, it is no more of a fake than the language itself. By the media's repeated proclamation that this creature is to be a "bear" time and time again, it becomes a bear. The language and the meaning of the language merge and become indistinguishable. However, to confess to the countries of America and Las Vegas that the bear is not a bear, the bear is just an imitation, a superreal figure, would be to destroy the expectation of reality itself. Reality demands consistency, and to destroy that consistency by admitting to forgery, whether deliberate or not, deconstructs the ideology of an entire society: "It is no longer a question of a false representation of reality (ideology) but of concealing the fact that the real is no longer real, and thus of saving the reality principle" (Baudrillard 13). The contest itself does not matter, nor do the contestants: "The public is completely uninterested in whether the contest is rigged or not, and rightly so; it abandons itself to the primary virtue of the spectacle, which is to abolish all motives and all consequences: what matters is not what it thinks but what it sees" (Barthes 15). In other words, to tell Curtis Norman that the bear is a meaningless robot would destroy his reality; to him, the bear looks real and thus is real, and there is no purpose served in destroying his reality by suggesting otherwise.

All of the books use nature as a contrast to the societal world in which the family typically exists. In *Bear v. Shark,* the "bear" and "shark" themselves exist as representatives of nature, in the same way that a family might think of a bear when planning for a family picnic, or may think of a shark when on the beach. In other words, if nature, in this sense aggressive nature, can be defeated, then the family will be preserved. The binary set up in this instance is society/nature, in which society is privileged because of its insistency; society, Mr. Norman learns, is smothering, in its very presence, but even more so in its inescapability. Traditionally, nature would be set up as a foil to society in that a family would physically make a journey into nature, as in two movies from 1975, *Jaws* and *Wilderness Family*. In these cases, the parents and children retreat into nature, only to find that nature is dangerous to the stability of the family unit; the task then becomes to defeat a natural element (a bear or a shark) in order to reassert the family's dominance over nature *and* over society (Morse). *Bear v. Shark* plays with this notion by making the Normans interact with nature, or as close as the Normans will ever come to nature: a simulated bear and shark, advertised and created within the machinations of a larger society. No matter how the Normans travel, they will never escape technology; nature no longer exists for them to retreat into: "The only phantasmagoria in the imaginary world lies in the tenderness and warmth of the crowd and in the sufficient and excessive number of gadgets necessary to create the multitudinous effect" (Baudrillard 12). Rather, the trip itself becomes a societal construct and a deadly one. At one point, Mr. Norman reminisces about a nature that he cannot even describe: "They all need to go somewhere pure, fresh, real. A waterfall, a meadow, a snow-capped mountain. Disneyland, with rides and well-scrubbed youths in costumes and cotton candy and oh Christ no not Disneyland but somewhere else green and fresh with water" (Bachelder 157). Here Bachelder reverses the binary; nature becomes powerful in its absence. Although it is nearly indistinguishable from simulation (Bachelder's joke about Disneyland references Baudrillard's famous example of Simulation of the Third Order), nature is still superior to simulation. Or, "Disneyland exists in order to hide that it is a 'real' country, all of 'real' America that *is* Disneyland (a bit like prisons are there to hide that it is the social in its entirety, in its banal omnipresence, that is carceral" (Baudrillard 12). In this instance, Mr. Norman recognizes that Disneyland is false, but cannot recognize that his imaginings of a natural world are just as simulated. Nature is simulation, because it is a preconceived, superficial notion. As Barthes quips, "The picturesque is found any time the ground is uneven" (74). Mr. Norman believes that a meadow or a waterfall will fix his marriage, his children, his own feelings of loss and inadequacy, but in reality, what he is wishing for is not nature but a miracle. In its simplest form, a waterfall is just an extension of a river or a stream, falling from a high place to a low place. For Mr. Norman, a waterfall assumes all the meanings of purity, escape from society and truth. However, because the ideal of nature does *not* exist, much as the sitcom family does not exist, society will continue to be the privileged binary, until the Normans can learn to differentiate between nature and the simulation of nature.

Nature, particularly the waterfall, is also a treacherous symbol in *Hell*. Davis's repeated proposition that "something is wrong in the house" would seem to be easily solved—if something is wrong in the house, characters must look *outside* of the house, in the natural world, for example, in order to escape what is wrong (Davis 1). However, again and again, nature is proven to be not a refuge but a place of execution. For characters in the dollhouse, to escape into the outside world is just to escape into a larger, more complex house; this experience is mirrored with Edwina Moss and the characters in 1950s Philadelphia. The house traditionally is a structure in which the inhabitants exist separate and safe from both nature and society. However, if something is wrong in the house—and *Hell* suggests that something is wrong in *all* houses—then the house, nature and society will blend to ill effect. In fact, the desire to escape from the house into nature is faulty on an even deeper level of simulation than a societal simulation of nature. Nature itself is master simulator; survival of the fittest ensures that every creature, in camouflage, mating rituals and eating habits engages in some sort of deception, some type of simulation. Nature, in *Hell*, is restricted to one sector, the Wissahickon, a deliberate foil to society: "Whatever its name meant, the Wissahickon was the closest thing we had to wilderness in a world otherwise composed of newly poured sidewalks, lawns like welcome mats, zinnias, marigolds" (Davis 120). Yet concrete, grass, flowers are all components of nature; if nature can be repurposed for a superficial, aesthetic design, then it can also be repurposed to simulate a feeling of refuge or escape, as the Wissahickon is repurposed. However, when society merges with nature, the wilderness will become corrupted. This is what happens when Joy Harbison's dead body is discovered in the valley: "Sunlight poured through the tulip leaves, the crowns of the hemlocks. On the gray tabletop among the puzzle pieces a red stain, a yellow eye, a black hightop—many things previously hidden were preparing to reveal themselves with terrible clarity" (Davis 124). The marriage of the unnatural with the natural, rather than providing a refuge or an escape instead merely corrupts the natural world. When Edwina Moss returns to the wilderness (presumably to this same spot), there is a similar moment of corruption. While there, her dog, as well as her friends, and perhaps even herself are corrupted; nature is not a refuge but a distraction. Though the language leaves the method of corruption up for discussion, the final passage in this wilderness seems to suggest that the human brain itself, which no one can escape, which is in itself the greatest metaphor for the house, for society, in the way it operates and has the ability to deceive and simulate without even conscious demand—the human brain itself is what has been corrupted. And while the human brain must be equated as being one with society, with the constructs of simulation, it may also be equated with the natural world:

> [W]here did the idea of an infinitely divisible whole come from in the first place if not from the human brain, its infinite folds, infinite pathways, tracks and trails, surging rivers, tree-crowned hills and dark-green valleys, its grazing cows, rabid dogs, hay bales, iridescent fish, leaping on sun-pricked hooks, rotted skin, quick sparks, waterfalls, all of its glittering synaptic leaps. (Davis 156)

In this way, the plight comes full circle. One cannot escape into nature; regardless of whether something is wrong in the house, the house is where one must live. The house is both the simulation of society and the simulation of the natural world.

In *The Miss America Family,* nature is also a consuming force; characters seek out nature to affirm their memory and identity only to find it has become a destructive entity. Pixie Kitchy's father originally affirms his identity as a youth by jumping into the Kill van Kull with his hands tied, a key in his mouth, only to emerge victorious. Through this act, in which he manipulated nature to simulate suicide, Pixie's father simulates the defeat of death. When Pixie's father returns to the river, he finds that what he has mistaken for a real defeat of death and mortality was a simulation, and attempts to recreate, and thus reaffirm some kind of truth, in his moment of triumph. In this way, simulation can be life-affirming, *because* it is indistinguishable from the real, or, at least, the mind can be fooled into accepting the simulation as permanent: "We require a visible past, a visible continuum, a visible myth of origin, which reassures us about our end" (Baudrillard 10). For Pixie, nature as a myth of origin is a reoccurring concept. Pixie uses the concept of moths in order to simulate her explanation of memory, her own myth of origin: "My mind is still sometimes more like a damp yard at dusk, fireflies lifting up, lit by a hundred tiny tinderbox fires. I know that doesn't make sense but I can think the word *memory* and see a cloud of moths" (Baggott 32). And again, Pixie's mother manipulates nature to create a myth of origin for not only herself but her family in its entirety: "I could go all the way back, couldn't I? Back to the beginning, a fish, an egg glistening in a sac, the only memory us half fish have of heaven. Remember that we are half fish? I have told you that" (267). For Pixie's mother, this simulation of memory is a myth of origin which serves two purposes; to begin with, it provides that visible continuum, visible past which is necessary for a religious woman who refuses to believe in Christianity because she believes Mary ought to have told the angel Gabriel, "No thanks," when he requested that she have a virgin birth. Secondly, in her old age, it gives her a simulation of memory, when in fact she is sliding into dementia.

Even when characters are not trying to appropriate nature as an escape from society or to reaffirm memory, the authenticity of memory is a dilemma in the works of modern American authors like Baggott, Davis and Bachelder. Memory, particularly in childhood, is difficult to trust because it is subjective, based on a variety of combined factors. Even the memory of an event like a birthday party can be manipulated so that in the retelling another attendee of the same party would not recognize it: individual experience, the experiences of others, emotions, quality of hearing and sight and the various intersections and interactions which occur at various stages of life will all come together in different degrees, affecting the quality and authenticity of memory. Therefore, all memory is both real and simulated. The simulation occurs as the memory is formed and for years afterward as the memory is shaped and reshaped to suit the needs of the individual. It is real because, whether or not a particular event or even a span of years objectively happened the way an individual subjectively remembers it, the individual believes in those simulated memories and thus builds an identity around those events or series of events regardless. For most characters, this simulation of memory allows for a sense of nostalgia, so long as the memory is not probed too deeply: "When the real is no longer what it was, nostalgia assumes its full meaning" (Baudrillard 6). Nostalgia can only exist in the absence of the real; the real is often disturbing if not traumatic. However, when the real has been simulated in memory, an individual is free to create a happier, wistful picture of the past.

For the characters of *The Miss America Family*, particularly Pixie Kitchy, identity, like memory is constructed based on the superreal because events of the past are surrounded by deliberate and distinct networks of lies. For Pixie, the greatest lie of her own identity was always the myth of her own beauty:

> One summer while stepping out of the Bayonne Public Swimming Pool in my pale blue swimsuit, I looked up and everybody was watching me. And it was like I was given a gun, something powerful, and everybody started acting like I had a gun, like I was armed and I could kill them if I wanted. It makes strangers awfully nice to you. It wasn't real power, I would learn… (Baggott 38)

One of the lies of beauty is that it indicates a correlation-implies-causation scenario when, in fact, there is no way to distinguish what comes first—beauty or power. Pixie supposes that her beauty gave her power, but part of this supposition is the simulation of memory. Part of it also seems to be an effort to relate herself to her handsome father, again an attempt to blur the distinct and separate spheres of reality and simulacra. Pixie wants to believe in the power of the superreal, the beautiful that is more perfect, more enchanting, more *real* than that of a healthy, young girl on the threshold of adolescence; the superreal translates into empowerment. In order to perpetuate this idea of the superreal, and in an effort to achieve the superreal, sitcom family, Pixie devotes a good deal of her life to becoming Miss Bayonne and then Miss New Jersey in the Miss America pageant system. The beauty pageant system itself is an excellent example of the superreal; these are young women who have been modeled as the very pinnacle of natural beauty, which must be true, because they are given titles and sashes and crowns, are broadcasted live on air as the most beautiful women in the United States. Yet, how natural is beauty when it is cultivated through deliberate planning, when winning hinges on learning a single song for the accordion, an instrument chosen not for passion or for talent, but because of its simulation of wholesomeness? This act of empowering oneself not through beauty but through a simulation of beauty is similar to what Barthes means when he describes the Parisian striptease: "[Accoutrements and clothing] all aim at establishing the woman *right from the start* as an object in disguise. The end of the striptease is then no longer to drag into the light a hidden depth, but to signify, through the shedding of an incongruous and artificial clothing, nakedness as a *natural* vesture of women, which amounts in the end to regaining a perfectly chase state of flesh" (Barthes 85).

In other words, Pixie enters the world of the superreal, a Simulation of the Third Order, in order to fully immerse herself in simulation. In laymen's terms, if you can't beat them, join them. She gains the mythology of power through stripping herself not to the essentialness of the classical nude but to the superreality of the beauty contestant. This manipulation of the superreal, the parade of lies is enough in itself to create a completely simulated identity (assuming that there is a natural state of identity, free from simulation, even if this state exists only at birth). When Pixie learns that even her memories of childhood—of *before* her discovery of her beauty, which she then cultivates into a simulation—are lies as well, particularly the lie of her father's rape, she loses perhaps the last "real" element of her identity: her prepubescent experiences. In light of this, Pixie engages in a search and destroy mission in which she attempts to destroy everything simulated in her life, beginning with her husband, for whom she has simulated love, affection, and obedience in exchange for the simulation of financial and emotional stability. This makes sense because, "Simulation is infinitely more dangerous [than transgression and violence] because it always leaves open to supposition that, above and beyond its object, *law and order themselves might be nothing but simulation*" (Baudrillard 20, original emphasis). When the truth of simulation is revealed, then, transgression in violence is a natural response to the destruction of an ideological order. Pixie's daughter is spared because of her utter *authenticity*—although Mitzie is preoccupied with appearances, pleasing people, she is obsessed in a wholesomely genuine manner; thus, her goodwill and enthusiasm is not a simulation but a state of truth for her. Ezra is mostly spared because he is perhaps the least ideal and thus least simulated aspect of Pixie's adult life; Pixie did not ask for a sickly son, yet still values him because he needs her. In addition, Pixie seems aware of her overdependence on Ezra. Finally, Pixie's mother escapes destruction because of her stroke. The stroke, while indirectly sending Pixie into an inspection of the simulated aspects of her memory and childhood, also creates a more authentic mother for

Pixie, one who has no need for deception. Next to Dilworth Stocker, the individual who suffers most from her destruction of simulation is herself. Without simulation in place, Pixie must develop a new identity; however, now she will be aware that any identity she constructs is only a simulation.

For Mr. Norman of *Bear v. Shark,* memory is indistinct because the flood of information through the media has made the authentic and the inauthentic indivisible; thus, his identity is also at jeopardy. The world in which Mr. Norman lives perpetuates the myth that anyone can participate equally in the world of adulthood. Curtis pretends to be Omar, engages in adult, eloquent, politically philosophical conversations of which his parents do not seem capable of understanding. Meanwhile, Mr. Norman cannot seem to understand how to be a parent or how to be an adult, partially because he cannot orient himself in his own memory of childhood. Mr. Norman confesses this fear that of his own disorientation to his wife:

> 'I mean, those people, us, that we see in pictures are younger, fresher, they have better shins. It's hard to identify with them.' […] 'But the scary thing is that it's not just my childhood. I have trouble remembering anything about my life, even from a week ago.' […] 'And then I think that maybe it's best that I don't remember. Maybe the truly scary thing is that there's really nothing much to remember.' (Bachelder 137).

Mr. Norman identifies a main problem of simulation; because memory is empty and subjective, individuals tend to fill in their own identities arbitrarily. But an arbitrary identity, based on arbitrary memories, is no identity at all. The identity will deconstruct under the slightest pressure; simulated identity needs the weight of belief, whether simulated or authentic, in order to continue its existence. Mr. Norman now is stuck in a world of childish pursuits, in which adults try to recreate their childhoods (and thus solidify new, stronger identities) by cheering in nonsensical debates or playing with technological gadgets: "The world wants to be childish in order to make us believe that the adults are everywhere, in the 'real' world, and to conceal the fact that true childishness is everywhere" (Baudrillard 13). As is also true for Pixie Kitchy, Mr. Norman's reexamination of his childhood only leads to the destruction of his loved ones. In his search for identity, Mr. Norman chooses to identify with a group, whose social identity is necessarily preconceived. In doing so, Curtis becomes brain damaged, ruining the only realistic hope Mr. Norman might have had for developing a healthy individual identity, as opposed to his unhealthy group identity, with the tools that were available to him, his own children: "Mr. Norman carrying ragdoll Curtis along the streets of Las Vegas. Mr. Norman can't figure out if he feels, carrying his young boy in his arms, like a good father or a very bad father. Mostly bad" (Bachelder 210). In effect, Mr. Norman does begin to develop an identity, as a poor father, a mediocre husband, and an overall discontented human being; his awakening to the simulation of his memory and identity is not a pleasant one.

The use of simulation, deconstruction and various elements of *mythologie* in these examples of modern American literature is mostly to the effect that all of these elements are a destructive part of American culture and society (and, more poignantly, all thought, regardless of nationality) particularly because of the way America disseminates information. Though these elements are destructive, it is also never beneficial to obsess over the superreality of modern American life. Overall, this sampling of modern American literature shows the tendency of this wave of literature to juxtapose its troubled families with the sitcom, ideal families of media and television, not because it yearns for the superreality, but because literature recognizes a disturbance in a lack of reality. There is a fear of simulation that is reflected in American literature: "The fear of simulation, then, is a concern that the outward appearances do not correspond to inner essences, and it [the fear] generates strategies to distinguish the apparent from the genuine, simulations from representations" (Dolan 63). Literature's purpose in highlighting simulation, the mythology of existence, the deconstruction of truth, is manifold. It exists as a coping mechanism, in which families who have been disappointed by the expectations of simulation can recover and find solace in others who experience and are manifestations of their trauma. Literature also exists as an instruction manual, a way of showing American readers how to separate the real from the simulation, the truth from the mythology, and the solid binary from the unstable binary. The problem, of course, in using theories like deconstruction and simulation in a discussion of American culture or society is that no one element can be privileged; there is no binary of authentic/simulation, reality/mythology. Instead, there is only information, and meaning to be gleaned from such information. Readers of novels like *Hell, Bear v. Shark* and *The Miss America Family* thus are armed with not just a fear of simulation, which has become ritualized over the course of the last several decades, but also with models of what awaits those who attempt to blindly ignore or else obsess over the concerns of those French philosophers; this literature also provides the modern American reader with the advice that they, Barthes, Baudrillard and Derrida, in their family of philosophy, might share around a kitchen table.

References

Bachelder, Chris. *Bear v. Shark: The Novel.* New York: Scribner, 2001.

Baggott, Julianna. *The Miss America Family: A Novel.* New York: Washington Square Press, 2002.

Barthes, Roland. *Mythologies.* Trans. Annette Lavers. New York: Hill and Wang, 1972.

Baudrillard, Jean. *Simulation and Simulacra.* Trans. Sheila Faria Glaser. Ann Arbor: U of Michigan Press, 1994.

Davis, Kathryn. *Hell.* Boston: Back Bay Books, 1998.

Derrida, Jacques. *Of Grammatology: Corrected Edition.* Trans. Gayatri Chakravorty Spivak. Baltimore: The Johns Hopkins U Press, 1974.

Dolan, Frederick M. "Fear of Simulation: Life, Death, and Democracy in Postwar America." *The Massachusetts Review.* 32.1. (1991): 61-75.

Morse, Kathryn. "Dad (And Mom) vs. Nature, 1975." *Environmental History.* 12.2. 346-49.

CHAPTER TEN

CHRIS BACHELDER'S CONSUMERS AND CORPORATIONS: *BEAR V. SHARK* AND THE PRICE OF TECHNOLOGY

JOE BAUMANN

American society, as suggested by Chris Bachelder, has become a mob of consumer zombies as a result of the world-wide investment in the technology that surrounds us. Human interaction, in fact, has transformed from primarily face-to-face contact into communication through technology and about technology, and Bachelder's *Bear v. Shark*, published first in 2001, is a frightening harbinger, a prophetic prediction of the future of American culture and of the current devolution of the ways in which we interact with one another. Bachelder's message of the degradation of American society through human technological consumerism, self-identification through possessions, and the loss of individuality at the hands of corporate monopolization are personified through the protagonist family, the Normans, and their interaction and daily living as they journey toward Las Vegas and the novel's titular death match.

Before getting into the specifics of the Normans' experience, it seems worthwhile to note the significance of their surname: the Normans are the normal, typical picture of the postmodern family in the world of *Bear v. Shark*, evidenced by their name, and the experiences that Mr. and Mrs. Norman (their first names are never given) and sons Matthew and Curtis have are the norm of the American family. Although the circumstances of their trip are unique thanks to Curtis winning an essay contest with thousands, if not millions, of entries, by naming them the *Norm*ans, Bachelder makes it clear that their daily life, with its heavy reliance on technology not only to communicate but to simply exist, as well as its revolving around Bear v. Shark II, is absolutely not unusual, and is, in fact, how every other family in this America operates. If their name is not evidence enough, the narrator at one point goes so far as to refer to the Normans as "the American family" when, in chapter forty-six, he says that "The American family just keeps driving the American vehicle across the American interstate system" (104): at this moment, if it has not been made clear already, Bachelder is certainly setting the Normans up as the iconic American family of this *Bear v. Shark* world.

As a result, the first morning we see at the Normans' home can be viewed as the typical morning for the *Bear v. Shark* family. A few points are of interest in the first few chapters of the novel. First, there is the matter of the television. Television, I would argue, is iconic of the technology takeover that has occurred in the America of the novel. As the Normans' morning begins, Bachelder describes the house as silent, "except for the chemical hum of the central air-conditioning, except for the hiss of dehumidifiers, except for the Babelic chatter of Televisions" (18). These televisions permeate everything in the Normans' life, and they cannot even be turned off, as evidenced in chapter five, which is comprised of a single sentence stating, "The old Televisions had an off switch" (22). They appear throughout novel, everywhere: not only is the one in the Normans' living room humming in the background as Mr. Norman slowly wakes up, another one awaits him on the stairs, and they cover the walls above each booth of the restaurant they stop at on their way to Las Vegas. These televisions are even in-tune with what a viewer wants to watch "the ESP TV reads your mind. From a comprehensive main menu, you program up to 38 General Topics of Personal Interest…and [the television] shows you what you want to see…Today's TVs flip themselves, move on, find the ideal programming and find it fast" (58).

These televisions are iconic of the technology that has pervaded the postmodern American identity as envisioned by Bachelder, and the power they hold over the characters in this world is frightening. Not only do these televisions—and, by extensions, other forms of contemporary technology—pervade the Normans' lives by being constantly turned on, they have changed the way people think and see the world. For example, in chapter eleven, when the Normans are attempting to figure out what the weather is like, Mr. Norman first checks the internet, but then wants to also check the weather on the television. When he asks Matthew to turn on the weather channel, a lengthy exchange about the variety of weather channels available (such as the Weather Network, Weather Network Plus, Extreme Weather, and Weather Europe) occurs, one that takes longer than it would to simply *walk outside* to see what the weather is like. This is clearly a two-fold commentary on Bachelder's part: not only is it a send-up of the egregious nature of the way technology has blossomed in terms of its supersaturation throughout American life (as seen in the overabundance of networks that focus on the weather), but it also indicates how reliant this family—and thus all of the America of *Bear v. Shark*—has become on technology for daily existence. While the family could simply go outside to see what the weather is like, they instead spend excessive time determining how to look up said information via a technological outlet. It would appear, here, that the American family does not know *how* to exist without technological support. We see this again and again at other points in the novel, such as when Mrs. Norman, instead of knitting, plays an electronic game simulating knitting. Thus technology has pervaded even the domestic minutiae of everyday life as well.

The morning conversation between Mr. and Mrs. Norman that precedes the discussion about checking the weather captures this nicely. When Mrs. Norman asks her husband how he slept, the following exchange occurs:

> Mr. Norman says, "There was an interesting program last night on the Great Wall of China. Turns out it's technically a hologram. It's the largest man-made hologram."
>
> Mrs. Norman says, "What do you mean by technically?"
>
> Mr. Norman says, "You mind if I turn up the birds?"
>
> Mrs. Norman says, "I dreamed that I found a turtle and I knew I had a CD stuck in him but I couldn't figure out how to get it out. It was so frustrating because I knew the CD was inside there, it was the Mall Sonatas one I like so much. Inside the turtle." (33-34)

This snippet of conversation demonstrates just how much technology has overrun the world of *Bear v. Shark*. Not only can technology control the bird noises that surround them (what these bird noises are, whether the sound of mechanical/holographic pets kept in cages or the sounds of birds populating trees outside is left ambiguous and unexplained by Bachelder) and thus replace the natural environment around them, technology also appears to have become capable of replacing famous icons; the Great Wall of China has been replaced by a hologram (or, perhaps, was one all along). Of note, however, is the fact that Mr. Norman had to be told this by a television program. Not only does this demonstrate the complete control television appears to have over the characters (Mr. Norman seems to be willing to believe this fact about the Great Wall of China without any feelings of disbelief or doubt, presumably because it has been provided by his television), but it is an example of the characters' growing inability to distinguish between the real and the created, a distinction that, as the novel progresses, is demonstrated to be one that has certainly begun to fade in this postmodern American world. Mr. Norman's job, as well, demonstrates this blurred vision:

> He works on a team that designs fake electronic equipment for model apartments and town homes…They just do fake electronic equipment. For model apartments and town homes. Well, it started out for model apartments and town homes, but lately regular folks have been buying the fake equipment because it looked better than the real equipment and it is competitively priced. (25-26)

This interest in fake equipment—equipment that looks better or more impressive than the real thing—is yet another example of the ways in which technology is invading and changing so many aspects of everyday life in this American landscape. Not only is technology replacing birds and other natural phenomena, technology is replacing itself with false versions of itself. As a result of their inability to identify things that are real versus things that are created, the population of Bachelder's America are easily led to want things that appear more real than things that are, in fact, real because, I would suggest, their entire conception of what is real has been radically warped, their ability to—or interest in—distinguishing the two rapidly declining toward nonexistence. As will be discussed shortly, this is something of a foreshadowing and reference to the fact that the fights between the bear and shark themselves are not real, either.

However, Mrs. Norman's recounting of her dream is worth a brief note for its significance to a variety of *Bear v. Shark*'s themes. First, it demonstrates just how much technology's hold on people has permeated their lives: not only does it consume their waking hours, but even peoples' dreams are saturated with technology. Mrs. Norman's dream also symbolizes the way that the natural world has been violated by the way technology has taken over the American world: in her dream not only has technology effectively gutted nature from the inside out (the turtle has a compact disc trapped in it), but Mrs. Norman's frustration shows how this overwhelming saturation of technology has changed the way people view technology itself. Mrs. Norman does not find her dream bizarre, nor does she seem to have concern for the turtle's well-being; her irritation in the dream is a result of her inability to figure out how to get at the technology she so desires, particularly because it is a CD that she loves so much. This demonstrates, I would argue, that people populating the novel have become so concerned and attached to technology that it has warped and narrowed their sense of what is important to the point of casting off concern for anything else.

As a result of this pervasive invasion of technology, people in *Bear v. Shark* find their identity through objects, ownership of said objects, and consumerism, exemplified through the materialism and commercial goods created in anticipation of the impending fight between the bear and shark. In fact, one could go so far as to say that the characters' identities within Bachelder's novel revolve exclusively around who they think will win, and any attempts or demonstrations of apathy about or distaste for the arguments about Bear v. Shark mark one as an outsider.

Take, for example, Matthew Norman. Matthew, in heavy contrast with his brother Curtis, demonstrates little interest in the question of who will win the fight between the bear and the shark, and when he does involve himself in conversations about the battle, his comments are often pithy: he asks, for example, "How many bears does it take to screw in a lightbulb" and constantly undermines these discussions with his antagonistic commentary. Despite his snark and sarcasm, however, Matthew is still susceptible to the pervasive powers of the Bear v. Shark phenomenon. When readers first see him, he is wearing clothing literally covered in the fight's paraphernalia: "Colorful sharks circle his pajamas and lunge at fat lazy seals." However, he has "cut off the sleeves, apparently with dull scissors or a knife" (34), an act of vandalism that can certainly be read as his way of expressing rebellion against this constructed, controlling power.

Despite this, I would argue that Matthew is in some ways one of the sole remaining voices of reason, as he appears to be the only one who questions the technological influences in the world around him. For example, when stopped for lunch, the Normans encounter a man at a restaurant who claims to have read a cockfight website that claimed that, in Bear v. Shark I, the bear threw the shark into the audience. This man clearly believes this, but Matthew points out that, "the same website also said that the bear shouted, 'I vanquish thee,' as he threw the shark which I doubt very seriously he did" (74). Unlike his father, Matthew at least appears to question the validity of that which is conveyed to him by surrounding technology.

This is one of very few occasions on which characters in *Bear v. Shark* ever question the things they are told by technology such as television and the internet, and the fact that Matthew is one of very few who does not buy into the hype and complete hold that Bear v. Shark has over the population as well as one who does this kind of questioning is hardly coincidental. The connective tissue here is the fact that the rest of these characters' identities have been subsumed to the side which they have taken in terms of supporting the bear or the shark, and, as a result, they are consumed by the consumables that they purchase as a show of which side they are on. Take, for example, Curtis: at the end of chapter twenty-eight, Bachelder writes that Curtis "had gotten chubby on Sea-n-Lea Meat Snacks" (71), a Bear v. Shark product. This small detail is an example of how a character is changed—here, even physically—by the phenomenon of Bear v. Shark.

One could argue, of course, that it is almost necessary for the people of this America to allow themselves to be engulfed in the Bear v. Shark phenomenon due to the cultural climate. Chapter seventy-three, "The Worst Kind of People," gives some insight into what happens to people who attempt to reject—or, at the very least, do not easily buy into without some thought—the bear and shark culture. While most people will "answer 'bear' or 'shark' very quickly and then proceed to provide reasons" when asked which side they are on, some people are "fence-sitters," though "there is very little apathy" (186) toward the question as a whole. These people are not the ones who are the worst, though:

> But there are some people, and you know the type, who refuse to answer until they have, say it with me now, *more information.* They say things like, 'Well what kind of shark are we talking about and what kind of bear? [...] And worst of all: 'Why would they ever fight? They have different ecological niches and they don't share the same food supply. A bear's digestive system blah blah blah. (186)

This passage demonstrates that those who cannot immediately make a decision about their position on Bear v. Shark are the most vilified in this fight-centered culture. Not only does the narrator say he hates that type of people, but even "Matthew hates these people, too" (186). This is not an insignificant detail. Matthew, as discussed, is arguably the only character in the novel that still retains some sense of rational perspective regarding the Bear v. Shark phenomena as well as some awareness of the deceptive powers of technology. Even he, though, cannot help but hate those who want more information, as well as those who question this cultural center of Bachelder's America. And it seems fair to say that people hate the people who question the premise of why a bear would ever fight with a shark: because the phenomenon has embedded itself into the entire world of the people inhabiting the novel, questioning the purpose of the fight itself (that is, its lack of a foundation in the real world) questions, in some ways, the culture these people are wrapped up in. This kind of questioning, and by extension rejecting, of all that Bear v. Shark stands for causes people to find themselves outcast from the world around them.

The novel and the characters in it seem aware of the need to join in the Bear v. Shark phenomenon. At one point, a radio talk-show host says:

> Join has undergone a revolution. Join has kept up. You have to keep up. If you don't, you get left in the dust. Join has reinvented itself. You have to hand it to join. Think of other verbs that have been left in the scrap heap of history, of interest only to crusty old professors...There is a technology of joining, Mitch, a whole technology. The satellite uplinks, the remote feeds, underground cables in networks and webs. We're all joined. (112)

Here, the novel makes a clear connection between being "joined"—or joining—and the technological developments that have made this joining possible; without the uplinks, feeds, networks and webs that have made instant connection possible, this joining would not be feasible. But there is certainly a negative connotation here: this joining, this adhering oneself to fads and trends, is something of a requirement in the America of *Bear v. Shark*; not only do people get "left in the dust" for not "joining," they are ostracized and labeled as the "worst kind of people," as suggested by the aforementioned passage on those who question the entire premise of the Bear v. Shark phenomena. Not only must one "join," that is all one must do: to ask questions, to ponder, to consider, is viewed in a wholly negative light. One must keep up but without dragging one's feet with questions. One must define oneself as a bear or shark supporter at the snap of a finger, with no less speed than it takes to plug into the technological network that allows one to broadcast one's position on the all-important question of bear or shark.

Characters not only become defined by whether they are bear or shark supporters, but their individuality is swallowed up by that choice and the consumerism that accompanies it. As noted earlier, Mr. and Mrs. Norman are never identified by first name, and rarely are other characters that the Normans encounter as they travel to Las Vegas given names, either. The aforementioned man with whom Matthew argues is simply "the guy in the booth next to the Normans," and even those opposed to Bear v. Shark II—those planning to bomb the event itself—are barely identified, called names such as Vice Squad Leader or Cult Leader. No one in this postmodern society fueled by Bear v. Shark and the consumer products and technology that surround them, it appears, is safe from this loss of individuality, even if one attempts, as the Cult Leader and Vice Squad Leader do, to reject Bear v. Shark and what it stands for.

Who, though, does Bachelder blame for this swallowing up of the individual identity and technology's commandeering of human thought? Is it the individual, the Mr. Norman or Curtis Norman who does not reject this postmodern society? The fact that even those who are rising up in rebellion against the spectacle of bear and shark are still anonymous and ineffectual suggests this is not the case. In particular, when Mr. Norman does reach out to these men, the result is permanent physical and mental injury to his son; Mr. Norman's efforts to reject and react against the popular mentality proves disastrous, suggesting that the individual cannot overcome this mindset and culture, and that trying to do so will end badly, implying that the individual him—or herself—is not directly responsible for the destruction of the postmodern American identity.

Instead, Bachelder appears to be inviting an interpretation in which the destruction of the American identity has come at the hands of corporate moguls rather than the fight between bear and shark itself. HardCorp, a large, ominous organization, makes many appearances throughout *Bear v. Shark*, often involving large demands from smaller organizations and creating the multimillion dollar event of Bear v. Shark II. The company is certainly and clearly depicted as an all-powerful organization resembling a governmental authority and is at one point even described in terms of its approval rating, creating a parallel between HardCorp and the presidential office. By doing this, Bachelder makes HardCorp seem like it has governmental control and authority, and as the company that sponsored and created Bear v. Shark, its control over American culture is clear. As noted, the event has consumed the American persona, and thus its profits are expected to be enormous, which will only allow HardCorp to have even greater control over the American identity as it creates more buzz and products following the fight that Americans will consume and be consumed by.

What is particularly important regarding this point, as well as the discussion of the deterioration of the American identity, is the fact that both Bear v. Shark I and II, as briefly noted above, are not *real*, in that the bear and shark that fight are holographic images programmed by employees of HardCorp rather than real animals. This fact *should* completely change the way in which the millions invested in this fight's outcome see the event: rather than producing an actual answer to the question of who would win this fight, the result of Bear v. Shark II is something produced by a corporation based on the use of technology. Who wins a fight between a bear and a shark, then, is not based on what an actual bear or shark would do, but what computer programmers decide would happen.

No one in the novel, however, seems to notice this distinction, and this detail is of central importance to understanding what has happened to American life as a result of the takeover by HardCorp and technology. HardCorp's motto, "Making life more real for over thirty years" (172), is iconic of the way in which reality and illusion have become blurred in this postmodern world, and offers up an explanation for why, in the aforementioned early scene in the novel, the Normans rely upon the internet and television for information about the weather instead of looking outside at reality itself. As established, for the Normans and the American world of *Bear v. Shark*, reality and its representation have become intertwined, blurred and indistinguishable. I would argue, in fact, that the reality constructed by technology, as seen most iconically in the illusory fight between the bear and shark, has *become* reality for the typical American. HardCorp, then, has succeeded not only in making life "more real" for over thirty years, but in warping the human understanding of what reality is.

Ironically, even the employees at HardCorp, it would appear, have a hard time grasping the distinction between their invented, engineered fight and the real thing. In chapter forty, "Bear v. Shark I: An Insider's Story," Alex Reid, an employee at HardCorp, says in his ghostwritten memoir *Swimming with the Sharks: My Two Years at HardCorp*:

> People in my generation never dreamed that they'd see a real fight between a bear and a shark. There have always been cock fights, of course, but I'm talking about the real thing—computer-generated, three-dimensional projection. I'm talking about action so lifelike, so realistic, that it makes real bears and sharks look like cartoons. (93)

This passage is chock full of irony: Reid identifies the "real" thing as something computer generated, a three-dimensional projection, rather than even considering that an *actual* bear fighting a shark would be the realest thing. As in HardCorp's motto, and as seen the Normans' daily life, Alex's conception of what it means for something to be "real" has become warped. This insight into the life of an employee of HardCorp also leads to an additional conclusion: that perhaps the individual employees working for HardCorp and working toward its aim of making life more real than real life are so far enveloped into the delusion that HardCorp as a whole has created makes it impossible to blame these individuals and their work for its effects, either, as they, too, have become victims themselves.

We see this trend in other small details as well, such as in a further description of Mr. Norman's work. As mentioned earlier, Mr. Norman works for a company that designs fake equipment for model homes. Of interest here is what the narrator tells us about their research for these projects:

> The design team knows nothing about electronic equipment except the way it looks. Team members scour gadget catalogs like porno mags. They have to keep up with technology. They have to keep up with the way technology looks. Team members e-mail each other when they have

> new ideas about how to make a fake piece of equipment look more real than a piece of equipment. (26)

This and other details throughout the narrative reveal the obsession with artifice that has overcome the world of *Bear v. Shark*, a consequence, I would argue, of the degeneration of the distinction between reality and invention in Bachelder's America. People, as evidenced by the popularity of Mr. Norman's company's work (everyone, as the success of HardCorp indicates, wants things that look more real than real life), are obsessed with the look of things. As a result, the disaster of Bear v. Shark I isn't that the fight doesn't happen or that it's a holographic image that isn't real, but the fact that the bear's head is too small and doesn't *look* how it should. HardCorp, in fact, fires the entire programming team that created the program because of the outcry the country responds to this "disaster" with, an indication that peoples' desire for correct artifice is pervasive and of highest priority.

What, then, is the price of this destruction? I would like to suggest that the answer to this question can be found in the structure of the novel itself. *Bear v. Shark* contains one hundred chapters in only two hundred fifty-nine pages, creating a work whose chapters are short, staccato, and, in some ways, disjointed. The suggestion I wish to make here is two-fold: first, that the way in which the novel communicates itself to the reader is reflective of the kind of disjointed communication that takes place throughout the novel. Chapters are often unrelated to one another in a literal sense, just as the conversations that occur between the Normans do not appear to follow a particularly coherent linear flow. Second, as the novel progresses, readers become accustomed to the brevity of the chapters, causing, essentially, a decrease in the requisite attention span a reader must have in order to engage the text itself. The result—in my experience, at least—is that chapter seventy, "The Ghost of the American Vacation," far and away the heftiest chapter at twenty pages, *feels* incredibly long, despite the fact that, in reality and relative to the chapters of many other novels, it is quite short.

Within said chapter, even, readers see further evidence of the effects of technology, consumerism, and the Bear v. Shark phenomenon. At one point, readers learn that "there was one Christmas a long time ago when Mr. and Mrs. Norman made each other gifts. It was a little rule they had. And the gifts were pretty good, too, and they meant a lot. Why did they stop doing that?" (174). The novel never answers this question, but the reason seems rather clear: their world, and the suffocating things around them, will not allow it. The *Bear v. Shark* America limits peoples' ability to take the time to do things on their own without technology's invasive hand; they have been so changed by the world they inhabit that they can no longer do things that they know used to mean so much to them. Just as Curtis has grown fat and Matthew has been disillusioned, so have Mr. and Mrs. Norman—and, by extension, everyone else in this America—been changed by its cultural climate and fads.

Finally, I would point out that Bachelder delays this long chapter and places it after so many short chapters to reflect onto the reader the declining mental acuity and stamina of the characters within the novel itself as a result of their break with reality and reliance upon technological advancement. Just as a reader who would otherwise normally have no problem reading twenty pages without a break might become "mentally fatigued" by the lengthy twenty-page chapter here, the inhabitants of the world of *Bear v. Shark* have become mentally fatigued by the technological and corporately-controlled world in which they live, and Bachelder shares this experience with the reader through the very structure of the novel. In fact, one could go so far as to say that the title of the novel itself reflects this clipped, shortened attentiveness and mental endurance: rather than calling the work *Bear Versus Shark*, the middle word itself is purposefully cut-off, creating a short-cut title, reflective of the shortened, lazy world that exists within the novel's pages.

Bachelder's *Bear v. Shark* is saturated with technology and demonstrations of its destructive effects. Not only do characters within this novel ignore reality and the tangible in favor of the electronic and technological, their identities have been completely absorbed in the worldwide phenomenon that is Bear v. Shark II. The subsuming of individuality to two large, impersonal identifiers—supporter of bear or supporter of shark—and the consumerism that accompanies this choice is identified, throughout the novel, as a destructive force as a result of the monopolization of materialism in American culture. Our obsession with technological advancement, Bachelder suggests, requires the sacrifice and price of individual identity and ability to distinguish reality from artifice, and bleakly argues that any resistance to this phenomenon-fueled progress is futile.

Work Cited

Bachelder, Chris. *Bear v. Shark: The Novel.* New York: Scribner, 2001.

CHAPTER ELEVEN

"NOT A MAN AMONG US CAN REMEMBER:" SOCIETAL WARNINGS OF *FRANKENSTEIN* IN *BUFFY, THE VAMPIRE SLAYER*

LAURA HOLDER

When Mary Shelley wrote *Frankenstein* in 1818, she began a tradition of using monsters as metaphors for different issues and problems within society. That tradition of monster as metaphor has continued into modern society, most notably in the realm of cult-television horror/sci-fi shows aimed at young audiences, such as *Supernatural*, *The Vampire Diaries*, *Teen Wolf* and, most notably, Joss Whedon's acclaimed series *Buffy, the Vampire Slayer*. For seven seasons of the show, monsters, vampires and demons act as thinly veiled metaphors for teenage angst and common issues of young adults. As the show progressed, however, the demons were used less and less to represent teenage drama and moved more towards larger social issues. Of specific importance for this paper is the main villain from the fourth season, a Frankensteinian type creature named Adam and the ways in which his existence and demise (like that of his allusive predecessor) questioned the ideas of (ab)normality and acceptance in society.

Specifically, *Frankenstein* and *Buffy, the Vampire Slayer* offer an insight into a panoptic society, what Michel Foucault describes in *Discipline and Punish: The Birth of the Prison* as a society of constant observation and enforced normalcy. According to Foucault, society underwent a major shift during the eighteenth century, setting it on the path of normalizing its citizenry and turning them into subjects that were as much under the scrutiny of their own society as they were in control of it. This new ideology worked to reproduce the basis of its power through the use of its institutions, both social and political. Foucault identifies the enclosed spaces of these institutions as a "place heterogeneous to all others and closed in upon itself" (141). Once everyone is classed and distanced from the rest of their peers inside these spaces, they have little to no choice but to become docile citizens because society has taught them they are reliant upon it, rather than themselves or each other. As such, they are leery of seeking conflict within the power structure that defines them. The idea of belonging to a communal body of society is replaced with the idea of the single individual, and anyone who appears interested in returning to that communal setting or speaking out in a communal voice becomes suspect and outcast from society.

As inmates within this social panopticon, the citizenry become part of the process of examination, a combination of hierarchy and judgment (Foucault 184). As a process, examination makes individuals into describable objects that can be analyzed and categorized into specific groups. Once that individual is placed into a category that is where he will stay. Examination works to "maintain him [the individual] in his individual feature, in his particular evolution, in his own aptitudes or abilities" (190). Maintaining the current behavior means maintaining current status; it is only by deviating from that behavior that an individual could be assigned to another category. In this way, the idea of "normality" is created. By specifying the details of everyday behavior, a sense of averageness, of normalcy, of acceptability came into being.

This "normalizing judgment" also worked to create another category – abnormality. By having a standard of normal to compare behavior to, abnormal or deviant behavior could now be recognized and properly punished (Dreyfus & Rabinow 158). Whether or not deviants found it within themselves to transform into the docile subjects that Foucault says panoptic society thrives on, they were assured of two things: they would always recognize themselves as deviants, and so would society.

***

ADAM: I'm a kinematically redundant, biomechanical demonoid designed by Maggie Walsh...

BUFFY: She pieced you together from parts of other demons.

ADAM: And man. And machine. Which tells me what I am, but not who I am. ("Goodbye Iowa" 4.14)

It might seem a stretch to say that *Buffy* has more than just a simple connection with Shelley's gothic novel, but many of the same questions and concerns are addressed by both. As Anita Rose notes in "Of Creatures and Creators: *Buffy* does *Frankenstein*," the novel looks at such issues as creation/creator dichotomies, acceptance, "the harm that disinterested science can do, and the arrogance inherent in an obsessive quest for knowledge" (134). These are all questions and concerns the twenty-first century must deal with, including the universe of *Buffy*. There are constant questions about responsibility and fault regarding the relationship of sires and the vampires they make. Is Angel responsible for the damage done by his "child" Drusilla, and therefore her child, Spike? The series also often plays on questions of acceptance – not only as a superhero who wants to be "just a girl," but of the value of demons. Do they have to be evil? Clem, an unspecified demon, is the Slayer's friend. Angel and Spike have their souls restored – can they be accepted even though they are still part demon? For the first three years, the show was centered in a high school, a location rampant with the constant desire to fit in, to belong.

The final two issues Rose brings up are where an obvious connection between *Buffy* and *Frankenstein* exists. During the fourth season, Buffy and company encounter two villains – the first is a paramilitary group containing both soldiers and scientists known as the Initiative. This federally funded organization works to capture, contain and study what it sees as demonic threats to human society. Dr. Maggie Walsh, a professor at the local college and one of the Initiative's top scientists, leads the way in this exploration. The Initiative has a very casual attitude towards the demons and creatures they capture and collect, often viewing them as nothing more than animals to be put in cages ("Doomed" 4.11). They do the occasional experiment, such as implanting chips into vampire's heads that cause severe neural pain as a prevention for them doing anything violent. Mostly, however, the Initiative embodies the idea of "disinterested science" Rose brings up. In the episode "Doomed," a demon escapes and must be recaptured. As a scientist tries to fill a syringe to drug the demon, a soldier, Riley, merely knocks it forcefully over the head and says, "Never mind" before walking off and continuing his conversation as if nothing happened. There is no concern about damaging the demon nor worry about what just happened. Just another day at work.

That disinterest, that arrogance or lack of view of possible consequences (not seeing the danger for the bright light of possible scientific discovery) is partially responsible for the creation of the main villain of the season, Adam, a creation purposefully reminiscent of Shelley's Monster. Pieced together from human, demon and machine remnants, Adam is created by Walsh as part of a secret project known as 314. Walsh calls the creature Adam, after the first man; similarly, Shelley called her Creature a "New Adam" (Wilcox 48). In both cases, these creatures are anomalies of authority, acceptance and belonging. They are each the only one of their kind, making them the absolute of authority and acceptance for themselves, and yet they are under the authority and considered abnormal by the societies of their creators. As the first of their kind, literal "Adams," both monsters have no category that they fit into.

Adam and the Monster start out as ideal societal subjects, something that Louis Althusser calls an "always-already subject" (1505). These always-already subjects are born blank with nothing more than the expectation to fulfill certain social familial roles. The Monster is like a newborn child, experiencing emotions and exploring his physical senses simultaneously while trying to form some type of understanding of the world into which he finds himself born (Shelley 95-6). As a newborn child, Frankenstein's Monster should be granted the right to carry his creator's name and assume an irreplaceable identity and space within Frankenstein's familial relationships and ideology. He is always-already granted this right, according to Althusser's theory. This subjection, however, relies upon the father being willing to take responsibility for the creation. But Frankenstein, the creator of this communal beast, is, in fact, the first to reject him.

Frankenstein sees the Monster and his first response is to run away and "escape" (Shelley 61). As creator and, essentially, father, in Althusser's terms, Frankenstein has a moral and social duty to the Monster. By not fulfilling that duty and not accepting his role as father and creator, Frankenstein denies his creation a place in society. As such, the Monster has no individual identity. He is simply viewed by society as Frankenstein ultimately views him, as a sum of parts that, while beautiful in their individuality, create a communal monster that has no one single, set place to fill in society. If, as Foucault posits, each "individual has his own place and each place its own individual," the Monster is automatically abandoned by society (143).

As an abandoned child left to fend for himself, the Monster should elicit pity from those he meets. However, most of society reacts to his appearance and presence as Frankenstein does. Frankenstein has no pity for the Monster, instead seeing demonism and horror in the Creature's first movements towards him. A weak smile and an outstretched arm, much like a child reaching for a parent upon first sight, are turned by Frankenstein into death grins and grasping hands to be escaped rather than embraced (Small 160). The rest of the Monster's encounters with society closely mirror this first meeting with his reluctant father. Those he meets either run away from him in fright or charge him in anger and fear, denying him entrance into society (Shelley 98). These individuals have been trained by society to recognize the Monster as an outsider, something unnatural and not like them, and therefore, something to be hated and excluded.

Where Frankenstein's Monster actively sought a connection with society and his father before ultimately rejecting both, Adam embraces the authority of his creator, Walsh. His manner of doing this, however, is not the normalized pattern of behavior. Adam's first act upon waking up to his new surroundings is not to reach out like a small child, but to stab Walsh through the heart. It is not that he doesn't know her or is reacting out of fear – he calls her "Mommy" as he releases her body to the ground. One might think Adam kills Walsh as an act of rebellion, but instead he sets about following the plans she had recorded on diskettes that he can read via his cyborg appliances. Much like how the Monster works to integrate himself into society – observing, learning to speak and read, patterning his behavior after others – Adam attempts to integrate himself into the role Walsh saw for him, forcing his way as an always-already subject in the name given to him by his "mother."

Since his rejection and abandonment by society can't come from his direct creator because he killed her, the rejection that sets him off to try and create a new society of his own must come from the next closest thing. For Adam, that rejection comes from Riley Finn, an Initiative soldier who was very close to Walsh. According to Adam's information, Riley was meant to be a part of the evolution of humans into chemically enhanced superior beings as designed by Walsh, a creation, like Adam:

> ADAM: She said that you and I were her favorite children. Her art. That makes us brothers. Family.
> RILEY: No!! I'm not like you.
> ADAM: That's pain, isn't it? ("Goodbye Iowa" 4.14)

Every time after that introduction, he does not call Riley by his given name, but forces a familial association by calling him "brother." The Monster was denied this opportunity by not having the ability of speech when he first woke up, so he could not force a claim of family upon Frankenstein. Who knows how the story might have changed if the Monster had been able to add a small cry of "Daddy" to his outstretched hand?

Riley rejects this familial bond, however, just as Frankenstein rejects his creation. One time, as Adam is detailing Walsh's plans for the two "brothers," Riley says: "Stop calling me that. I'm not your brother. You're a botched science experiment. I'm a human being" ("Primeval" 4.22). Riley is desperately trying to set himself apart from Adam, from the abnormality of it all, specifically identifying himself as human, as if there is some particular saving grace in that distinction. And to a certain extent, there is as far as social acceptance is concerned. Human is known, understood, categorized – race, ethnicity, gender, hair color, eye color. As Foucault defined society, we have sections for each and all, and because each has a section, that makes them normalized, not alone – not abnormal. Neither the Monster nor Adam can fit into any one category. They are amalgamations of all kinds and therefore belong to none. Not only are they outsiders, but they are monsters by society's standards. Evil. To be destroyed, at all costs.

Like the Monster, Adam does not start out as primarily evil, however. Though his actions are viewed by human society as evil and monstrous, to him they are logical and non-threatening. Adam possesses an almost child-like sense of wonder and curiosity and, even when he is at his most heinous, calls events merely "interesting," because, to him, they are merely experiments, not actions of malicious intent. The Monster is likewise cast as a child who knows nothing of his world and must explore it to learn the ways that it works.

The perceptions of Adam as a monster depend solely on the background of the characters that are viewing him, which is where Whedon begins to play with the idea of society's responsibility towards its treatment and expectations of individuals outside society's accepted view of normal. When Adam first ventures into the outside human world, he is unaware of what he is until he encounters a human presence, just like Frankenstein's Monster:

> ADAM: What am I?
> KID: You're a monster.
> ADAM: I thought so. What are you?
> KID: Me? I'm a boy.
> ADAM: A boy. How do you work?
> KID: I don't know. I just do. ("Goodbye Iowa" 4.14)

The boy doesn't even stop to consider that Adam might not be a monster because society has already taught him what monsters look like. This is in part a reference to the prominent place Shelley's *Frankenstein* holds in our society as well as to the fact that it has set the bar of consideration for all monsters. Because we have seen Frankenstein's Monster so much, we have assimilated it into our understanding of normalcy and monstrous. *Frankenstein* has become what Frederic Jameson calls an "always-already-read" text, meaning we always approach it with an "already" set, accepted reading and interpretation, even when we're looking at a new reading or criticism of the work (1937). Thanks to the prevalence of monsters, specifically Shelley's Monster, in modern society, the small boy Adam meets automatically categorizes Adam as a monster – something foreign and alien and outside of acceptable human society.

The Monster in *Frankenstein* is also categorized as such by an outside source. When talking to Frankenstein, he reveals that he learned of his own personal deformities by observing others in society and then looking at himself: "I had admired the perfect forms of my cottagers ... but how was I terrified, when I viewed myself in a transparent pool!" (Shelley 104). Until this point, the Monster had been keeping himself hidden not because he knew he was monstrous but because of his past experiences with individuals in society. But what prompts his recognition of his monstrosity? Unlike Adam and the boy, he has no iconic cultural reference to compare himself to. For Frankenstein's Monster, his recognition comes from Foucault's social institutions – namely education – that are responsible for identifying and classifying normal and abnormal, human and monstrous.

The Monster did not understand the differences between himself and those he observed until after he receives the proper educational training, i.e. learning to speak, read and write. In his article "'Master of Their Language': Education and Exile in Mary Shelley's *Frankenstein*," John Bugg elaborates on Shelley's connection between education and the Monster's realization of his own difference from the De Lacey family: "Because the Creature has this realization [about his monstrosity] directly after the passage on education, it seems that he has learned both language and the very system that makes him monstrous" (661). Rather than seeing his reflection and calling the cottagers monsters and deviants, the Monster's education has prepared and trained him to recognize and accept himself as the true monster, one who could never be accepted in society.

The perception of monstrosity, however, is confronted in *Buffy* when Adam meets other demons, namely vampires. These beings, who themselves are hybrids of demons and humans, who have both demonic and human appearance, recognize and accept Adam as something very similar to themselves because he presents himself as a, "Demon in a human body. You walk in both worlds and belong to neither. I can relate" ("Who Are You?" 4.16). In fact, Adam becomes the bridge between vampires and true demons, two groups that had previously been opposed to one another, and brings them together on the same side. From this perception, Adam is not a monster – he is a diplomat who is working to create a society in which all breeds of demons and hybrids can interact on a peaceable level.

But Adam is monstrous, simply because he is not being viewed by demons or hybrids, but by human society. In the same way that most readers feel pity or sympathy for Frankenstein's Monster, we can view Adam as something other than being ONLY a monster, but they are both still monsters, nonetheless. They are the antagonists to be railed against because they are outside of society. Which is why both the Monster and Adam abandon traditional society in favor of creating their own groups. The Monster, having fully accepted the fact that he is never to be allowed entrance to Frankenstein's society, orders his maker to create a female mate for him. He declares that his "companion must be of the same species, and have the same defects" as himself, therefore separating himself from human society even more (Shelley 128). In much the same way, Adam sets about creating his own society of demon-human-cyborg hybrids. Though Adam is not specifically looking for a mate, he is looking for a group, a society in which he is not abnormal.

Frankenstein refuses to finish the female monster because he fears the possibility of monstrous children coming from the union. He fears the creation of an abnormal society that could threaten the existence of his own normalcy. Adam's society is also not seen as desirable by the human protagonists of the show and, as such, the human audience. It would require the deaths of humans and demons alike, from which would be born more versions of Adam. He is willing to force their creation on the world just as the Monster attempted to force Frankenstein to create a female mate for him. Adam views this forced collaboration as the beginnings of a utopia in which society will not be able to discriminate individuals based upon birth, class, race or creed. All individuals will be comprised of all areas of society – to discriminate against one would be to discriminate against all.

***

> *Adam: You feel smothered. Trapped like an animal, pure in its ferocity, unable to actualize the urges within... clinging to one truth like a flame struggling to burn within an enclosed glass... that a beast this powerful cannot be contained. Inevitably it will break free and savage the land again... I will make you whole again. Make you savage. ("The Yoko Factor" 4.21)*

Where *Buffy* and *Frankenstein* part company is in the demise of their monsters. Frankenstein's Monster elects to destroy himself after the death of his creator. His death is of his own choice, and in that he regains some authority and agency for himself in society. He plans to burn himself so that his "remains may afford no light" to anyone who attempts to re-create another such as himself (Shelley 188). His final words work to spread himself out into society more than his original existence ever could because, rather than presenting an image of himself as a regretful outcast, he claims he will leave this world "triumphantly": "My ashes will be swept into the sea by the winds, and my spirit will sleep in peace" (189). The spreading of his ashes out to sea represents the presence of his communal and outsider ideology that will continue to exist in society, in fact will continue to be created by society, just as he himself was created by Frankenstein. Though he is lost to the "darkness and distance" in the final lines of the novel, the Monster has made himself immortal in the realms of society as the ultimate outcast.

Adam does not go triumphantly, or even of his own accord. This may be seen as the difference his acceptance of Walsh (even though he killed her, he still followed her plan) made as opposed to the Monster's disassociation with Frankenstein. Adam does not see himself as a complete outsider with no hope of ever rejoining society. He recognizes himself as a monster, yes, but instead plans to renovate society so that it contains monsters. He is still searching for a community that will accept his communal self, even if he has to create one from the scraps of the existent society.

Where the Monster's death shows the inevitability of trying to exist outside of society, Adam's existence and his defeat work as examples of an emergent post-modern society that deals in fragmented communities. In order to defeat Adam, as he must be defeated seeing as how he is the villain, the protagonists must perform a magical spell that joins their essences into one being. The knowledge of the watcher, Giles, the loyalty and heart of the true friend, Xander, and the magical power of the Wicca, Willow, are all joined together with the super-strength of the slayer, Buffy. It is only by becoming a similarly communal entity that the protagonists are able to defeat the villain, thus proving that the hybrid concept of Adam is not entirely evil. As with Frankenstein's Monster, the evil comes in immediately rejecting him, from not attempting to understand him.

Mainstream American society is drowning in a sea of concerns dealing with individuality, community, and outside identity, ranging from legal and illegal immigration to nationalizing a language to legalizing gay marriage and anything in between. We are in constant arguments about whether to repress these individuals and maintain the norm, the status quo, or try and broaden society to include those we see as deviant, as abnormal, and in the case of some extremist views, as "demons." We worry that if we allow them into our social boundaries, will they not then take over and destroy our balance of normality? Against this background, Adam acts as a similar warning, but not against letting them in. Adam is a warning of what might occur if we continue to keep them out: if modern society does not allow outside individuals to find their own place within mainstream society as opposed to being driven out or kept on the edges of that society, they may very well attempt a coup or rebellion against mainstream society. While it's highly unlikely that segregated groups will call on demonic forces to physically demolish the dominant society, there is nothing to say that they won't be more than happy to tear down the institutions that have kept them alienated and set aside as deviants and use the scraps to create their own accepting community.

In response to creating the communal force to defeat Adam, Buffy is visited by the essence of The First Slayer in a dream – a primal force, alone and individual. The First Slayer condemns Buffy's choice to have friends, to act as part of the composite with them and attacks her, saying "we [slayers] are alone" ("Restless" 4.23). This is seemingly contradictory, however, because The First Slayer has no voice. She has someone to speak for her – Tara, another of Buffy's friends. Tara claims she was "borrowed" for the dream, to speak for the First Slayer: "I have no speech. No name. … I am destruction. Absolute… alone" ("Restless"). After a brief struggle, Buffy stops and says, "It's over. We don't do this anymore." When the First Slayer attempts to continue the fight, Buffy screams, "ENOUGH!" and wakes up ("Restless"). When Giles attempts to explain the dream (which all four participated in, much like their communal joining against Adam), he says, "Somehow our joining with Buffy and invoking the essence of the Slayer's power was an affront to the source of that power" ("Restless").

In this same dream, Buffy also encounters the original human component that was Adam. She asks what his name was and he replies: "Before Adam? Not a man among us can remember" ("Restless" 4.23). Adam is timeless, harkening back to the first man and to Shelley's original composite monster. This relationship connects not only Adam, but the entire gothic genre of Whedon's series, back to its roots as well as connecting the audience back to the idea that this fear of the Other, the unknown, is nothing new. Because the Monster's death is not actually shown within the pages of the novel, *Frankenstein* never truly ends (Small 195). The readers are left in the "darkness and distance" of the Arctic Circle, waiting to see the light from the Monster's funeral pyre (Shelley 189). Likewise, the ending of Adam, the defeat of a composite monster by the creation of a similar composite, offers the notion that Adam's idea of a community of hybrids might not be too far away.

References

Althusser, Louis. *Ideology and Ideological State Apparatuses*. 1970. *The Norton Anthology of Theory and Criticism*. Ed. Vincent B. Leitch. New York: W.W. Norton & Company, 2001. 1483-1509.

Bugg, John. "'Master of Their Language:' Education and Exile in Mary Shelley's *Frankenstein*." *The Huntington Library Quarterly* 68.4. (2005): 655-68.

Dreyfus, Hubert L. and Paul Rabinow. *Michel Foucault: Beyond Structuralism and Hermeneutics*. U of Chicago P, 1982.

Foucault, Michel. *Discipline and Punish: The Birth of the Prison*. Trans. A.M. Sheridan. New York: Pantheon Books, 1977.

Jameson, Fredric. *The Political Unconscious: Narrative as a Socially Symbolic Act*. 1981. *The Norton Anthology of Theory and Criticism*, Ed. Vincent B. Leitch. New York: W.W. Norton & Company, 2001. 1937-1960.

Rose, Anita. "Of Creatures and Creators: *Buffy* does *Frankenstein*." *Fighting the Forces: What's at Stake in* Buffy the Vampire Slayer. Eds. Rhonda V. Wilcox and David Lavery. Lanham, MD: Rowman & Littlefield Publishers, Inc., 2002. 133-142.

Shelley, Mary. *Frankenstein*, Ed. Johanna M. Smith. Boston: Bedford/St. Martin's, 2000.

Small, Christopher. *Mary Shelley's* Frankenstein*: Tracing the Myth*. U of Pittsburgh P, 1973

Wilcox, Rhonda. *Why Buffy Matters: The Art of Buffy the Vampire Slayer*, New York: I. B. Taurus, 2005.